To Marley

Happy B-day Month
I love this book and read
a little each day – _savoring_ it!

SACRED PLANT MEDICINE

I know I purchased a book
for you by this Author
this is the newbook by Buhner

Hope you love it like
I do ... I am on page 70 - its great!

Love Always
Laurie

Also by Stephen Harrod Buhner

THE TASTE OF WILD WATER
Poems and Stories Found while Walking in Woods

THE SECRET TEACHINGS OF PLANTS
The Intelligence of the Heart in the Direct Perception of Nature

THE FASTING PATH
The Way to Spiritual, Emotional, and
Physical Healing and Renewal

VITAL MAN
Natural Healthcare for Men at Midlife

THE LOST LANGUAGE OF PLANTS
The Ecological Importance of Plant Medicines
for Life on Earth

HERBS FOR HEPATITIS C AND THE LIVER

HERBAL ANTIBIOTICS
Natural Alternatives for Drug-resistant Bacteria

SACRED AND HERBAL HEALING BEERS
The Secrets of Ancient Fermentation

ONE SPIRIT MANY PEOPLES
A Manifesto for Earth Spirituality

SACRED PLANT
MEDICINE

The Wisdom in Native American Herbalism

STEPHEN HARROD BUHNER

Foreword by Brooke Medicine Eagle

Bear & Company
Rochester, Vermont

Bear & Company
One Park Street
Rochester, Vermont 05767
www.InnerTraditions.com

Bear & Company is a division of Inner Traditions International

Note to the reader: This book is intended as an informational guide. The remedies, approaches,
and techniques described herein are meant to supplement, and not to be a substitute for,
professional medical care or treatment. They should not be used to treat a serious ailment
without prior consultation with a qualified health care professional.

Library of Congress Cataloging-in-Publication Data

Buhner, Stephen Harrod.
    Sacred plant medicine : the wisdom in Native American herbalism / Stephen Harrod Buhner;
foreword by Brooke Medicine Eagle.
        p. cm.
    Originally published: Boulder, Colo. : Roberts Rinehart, c1996.
    Includes bibliographical references and index.
    ISBN 1-59143-058-5
    1. Indians of North America—Ethnobotany. 2. Indians of North America—Medicine.
3. Traditional medicine—North America. 4. Medicinal plants—North America. 5. Plants—
Religious aspects. 6. Human-plant relationships—North America. 7. Hallucinogenic plants—
North America. 8. North America—Social life and customs. I. Title.
    E98.B7B74 2006
    615'.32108997—dc22

                                               2005034206

Printed and bound in the United States by Lake Book Manufacturing

10 9 8 7 6 5 4 3 2 1

Text design and layout by Rachel Goldenberg
This book was typeset in Sabon, with Alexa and Bauer Text Initials as the display typefaces.

To send correspondence to the author of this book, send a first class letter to the author c/o
Inner Traditions • Bear & Company, One Park Street, Rochester, VT 05767, and we will
forward the communication to the author.

See page 208 for a list of permissions to print photographs and to reprint extended quotations
from copyrighted works.

CONTENTS

For My Teachers

For Creator

For Grandmother Earth

For the Plants

ACKNOWLEDGMENTS

So many people affect a life and a work that it is not possible to acknowledge them all without a listing as interesting and ponderous as a phone book. But there are many people who helped in the birthing process of this book and I want to thank them here.

The medicine people who had the courage to share their understanding before they died, particularly Main'gans and Owl Woman; my great-grandparents, C. G. and Mary Harrod, for reasons that are clear in the book; my grandmother, Edna Buhner, for teaching me the meaning of friendship; my son, Benjamin, for his companionship; my wife, Trishuwa, for endless late-night conversations and suggestions and support; Brooke Medicine Eagle for her courage, endless devotion to her path, and visions of the pipe; Wabun Wind for her strength of character and many phone conversations of support and education; Ed Canda for 15 years of friendship, mentoring, and companionship as an ecstatic; Rick Rinehart and Toni Knapp at Roberts Rinehart for believing in the book; David Rockwell for his integrity and insight in *Giving Voice to Bear*; Cyncie Marini for her wonderful paintings; Catherine Hunziker for introducing me to the plants and thus beginning the journey; Kelly Kindscher and Bill Lyon for their early reviews and their writing; and John Dunning for friendship and wisdom and reading endless early drafts of my writing and not laughing or even smiling because he knew how easily the small flame of hope can sometimes be extinguished.

FOREWORD

By Brooke Medicine Eagle

*A*t a time when one of the most poignant issues for the family of two-leggeds is reacquaintance with the sacred, Stephen Harrod Buhner's book offers a wonderful sense of that journey. His own process of deepening that experience, which he shares throughout the book, and his discourses on the nature of the human quest for the sacred are profoundly useful to anyone who seeks wholeness and holiness. Whatever your specific mode of seeking knowledge of the sacred, this book will help you deepen it into greater knowledge and, more especially, into wisdom. For this book is not simply a book of facts, it is a book of experience, of knowledge deepened into wisdom. And like all wisdom, it teaches you not only about the specific subject (in this case the sacred use of plants) but also how to live a good life. It generates from the simplest things a profound understanding of sustainable, harmonious living. And this understanding, this knowing, is the most valuable thing in our world today.

Sacred Plant Medicine beautifully expresses an Earth wisdom frame of reference, not only in the extensive quotations from native elders but also in the way the information is given, the very way Stephen has approached his own learning. He makes it clear that one does not have to be of a certain color or race or ethnic background to seek and find a way of being in harmony with this North American land. It seems natural to seek this ancient herbal knowledge from native people, yet we must awaken to our full humanity and understand that spiritual Earth wisdom is everyone's birthright. We have only to devote our attention to its revelation. To find harmony, we *must* awaken to this ancient vibrant potentiality.

Stephen notes that, "It is difficult for people coming from a Western perspective to understand the mindset necessary to be in this kind of [sacred] relationship with plants." As you read through the book, you will find yourself steeped in that mindset, will find it seeping in through the cracks we have made between us and other lifeforms. You will joyfully awaken to these ways that are natural to all human beings. If you travel this road with Stephen long enough, you will feel something special, ancient, intrinsic, stirring inside of you.

Sacred Plant Medicine offers us ways of healing the greater body of the Earth by focusing our attention where the support for human life began—with the plant people. Every chapter is an eloquent expression of the almost magical power of truly listening to the life around us, of honoring the vital consciousness in all things, of deeply respecting the wisdom of all life. Our salvation obviously is not in the conquering of nature, as has been practiced now for generations with dire consequences, but in becoming "masters of the wilderness" as some have described the indigenous peoples who have through time understood fully how humans are woven into the web of life.

Most refreshing and empowering in what is expressed here is that we do not need advanced university degrees or to study with the highest gurus to heal ourselves and our planet. We need simply to learn the feminine lesson of opening ourselves to the life immediately around us, which will remind us to remember the deep knowledge we possess as a birthright.

Our prophecies have long told us that the real teachers of this new time will be the farmers and the people whose hands hold the life that sustains us. And so I was glad to read in Stephen's final chapter the words of Bill Mollison, who is one of my models for a good life. Bill is devoted to helping us rebuild a sustainable culture (Permaculture) through creating sustainable agriculture and lifeways. In his wonderfully outrageous way, he encourages us to read the "book" of nature and to use the rest of the books for compost. My elders have said to me that the trees are the teachers of the law, and I have long contemplated what this means. As I grow less ignorant, it is obvious that the true university is the forest, where a sustainable, rich life is evident to all who will engage this world directly and openly.

Not only the trees, but all the plant peoples, have much to teach us. They are our forebears on this sweet Earth, having made a hospitable place for us through their cycles of oxygenation and water. In addition to this material symbiosis, they have also integrated the emotional issues that we two-leggeds struggle with so often. By partaking of their vibratory intelligence through ingestion as well as through communion, we can become more whole and integrated ourselves, and thus extend a healing vibration to the life around us. For it is in the emotional healing of our relationship issues that the world will become whole.

As this egg of the new time cracks open, we must rebirth ourselves into lifeways that are in good relationship and will sustain all life over time. What we have to gain is not just the averting of disaster, but a renewed life of beauty, abundance, radiance, rich greenness, and bird song—a true gift to the children of seven generations. We will be happy as our mother Earth is happy. The Christian Bible tells us that the humble will inherit the earth; in these days of crisis, we can understand this quite easily when we realize that the word *humble* comes from the same root word as *humus* and means "of the earth, or close to the earth."

In *Buffalo Woman Comes Singing,* I relate the story of the coming of White Buffalo Calf Pipe Woman and the holy gifts she brings us through the sacred pipe. This pipe, which represents oneness and wholeness—a healing for the Earth—is one of our most potent ceremonies. Stephen tells of his loving relationship with the pipe and the many ways in which he has deepened his relationship with it. He clearly conveys the understanding of how sacred an act the use of the pipe is. It is not about "playing Indian" or impressing your circle. The pipe ceremony is for honoring and praying for All Your Relations, something that is vitally important in this time. The commitment to the pipe is to use it for healing and to be willing, at any time, to go wherever you are called to perform that ceremony. The pipe carrier is a servant of the people, committed to restoring the web of life.

White Buffalo Calf Pipe Woman reminds us of Creator's one law upon forming this mother Earth: You *shall be in good relationship with all things and all beings.* Her corollary to that law reminds us poignantly: *Whatever you do to any other thing or being in the circle of life, you do to yourself, for you are One.* Whether it is sacred plant medicine or with anything else you do, this understanding is important, and Stephen again and again gives us exercises and ways of making this real in our lives. His exercise of the inner medicine wheel is excellent, using an ancient ceremony within ourselves to come to wholeness. Then the work with the outer medicine wheel can be even more powerful. Again and again, he asks us to take the work within, to a feeling place, to come into more wholeness as a result of it. As well, he admonishes us to realize that our relationship with the plants is not just for ourselves. We must always go to sacred and powerful things or places or beings, willing to give of ourselves in service of them. Our approach should always be interactive, rather than expecting or demanding. It must be as it is with any true friend; we are there in mutuality and love. We must be willing to give of ourselves to help these friends, and this will very likely call us into action as sacred ecologists.

In summary, what is provided in this book is a way to restore our connection with all of life, to awaken ourselves to the beauty and the possibilities of life. In the process we will come to know ourselves and come to know our sweet earthly home with an intimacy and a joy unheard of now. What we must offer is our attention, our intent, and our patient listening to the great voices around and within us.

May the healing of the sacred plants be yours.

PREFACE
TO THE NEW EDITION

This book is about a particular way of gathering information from the world, not the reductionism that our modern culture so embraces, but an older way known to all ancient and indigenous cultures. It is a way of gathering information directly from the world itself, a way of learning the uses of plant medicines directly from the plants themselves.

Members of most ancient and indigenous cultures make an interesting assertion; when asked where in their body they live, they gesture to the region of the chest. Members of our culture, on the other hand, point to the head, generally an inch above the eyes and about two inches into the skull. The great divergence in the ways that Western and indigenous peoples experience the world can, I think, be traced to just this difference. For those locating themselves in the heart and those locating themselves in the brain do experience the world in quite different ways. Realms of experience, open to those who approach the world through the heart, are simply not perceivable to those who experience it through the brain.

Consciousness studies, begun in the late 1960s, focused almost entirely on the brain. The Western assumption has been that it is our brains that make us different from the other inhabitants of this Earth.

But in the last two decades a number of researchers have approached consciousness without that preconception, which has allowed them to be more open in their explorations. Some of them began to realize that consciousness is not limited to the brain, rather it is highly mobile, easily habituating itself to other locations in the body. So they began to look at the heart and its role in cognition and awareness. One of the most important recognitions emerging through recent studies of the heart is that our individual organs, as well as the entire human organism itself, are not linear expressions, but are highly complex nonlinear organisms in which the whole is far more than the sum of the parts.

Researchers have found that when billions and billions of molecules are placed in an enclosed container, their movements are, at first, random. But at some moment, that can never be predicted, all the molecules spontaneously synchronize. They begin to move and vibrate together and tightly couple with one another into a single coordinated whole. They become one system of which the individual molecules are subunits. In that moment of synchronicity, something comes into being that is more than the sum of its parts. And that something, the soul of us, can never be found in any of the parts, no matter how minutely they are examined. At that moment of synchronicity, the new system also begins to display what are called emergent behaviors. The whole system begins to act on the parts, or subunits, to stimulate further, more complex synchronizations. A continuous stream of very rapid information begins to flow from the parts to whole and whole to parts in order to strongly stabilize the system. This information stream goes to and from the environment as well in order to analyze any external factors that could affect the system's stability.

The information stream may occur in numerous languages: temperature fluctuations, changes in velocity or pressure, adjustments to chemical composition, the flow of electromagnetic signals, and more. But, of course, as with our verbal language, it is not the word itself that is important, but the meaning within it that is of the essence. The meanings within a molecule, its electromagnetic (EM) signature, tell a receiving organism how that molecule can affect its state of being. These meanings are analyzed, integrated into the organism, and a response is initiated.

All living systems are like this, they self-organize and show emergent behaviors. They all retain an exquisite sensitivity to perturbations of the equilibrium that they experienced at the point when they self-organized. In fact, they remember that moment of equilibrium and remain attuned to it throughout their lives. The threshold they crossed is itself a living identity to them. So they monitor their internal and external world through extremely tight couplings at billions upon billions of points of contact. In this way they are able to process all of the energy, matter, and information that comes to them in order to maintain their existence. In other words, they are highly intelligent and they all possess this soul force, this thing that comes into being that is more than the sum of the parts. The heart is just such a nonlinear system. It possesses both self-organization and emergent behaviors. It functions not only as a powerful endocrine gland, but also as a unique kind of brain, a cognitive and perceptual organ, and a powerful electromagnetic generator and receiver.

The heart contains what are called pacemaker cells, which set its regular beating pattern. At the moment of self-organization, the first pacemaker cell begins beating, pulsating or oscillating, at a regular rhythm. Every new pacemaker cell that forms hooks itself to this one, and the new cell begins beating in time with the first. If one of the heart's pacemaker cells is removed from the body, kept alive, and placed on a slide, it loses its regular beating pattern and starts to fibrillate, to beat wildly and irregularly, until it dies. But if you take another pacemaker cell and put it close to that one—they do not need to touch—their beating patterns will synchronize; they will beat in unison. A fibrillating cell, placed next to a non-fibrillating pacemaker cell, will stop fibrillating and entrain, or begin to beat in unison with it. The reason they do not need to physically touch is that they are producing, as all biological oscillators do, an electric field as they beat. Only the fields themselves need to touch.

Individual pacemaker cells tightly couple together by the millions in the heart. The field that they produce together is much larger than that created by each individual cell alone. It is 5000 times stronger than the brain's electromagnetic field and can be measured by the most sensitive scientific instruments up to ten feet from the body. The field is strongest within eighteen inches of the body surface but continues out indefinitely into space, traveling in the same way that radio waves travel. (You can get a sense of this field if you have a friend stand five or six feet away from you, with his or her arms stretched out wide to the sides. If you then walk up to your friend, slowly, you will, when you are about 12-18 inches away from them, suddenly experience yourself being "in their space." This is an experience of your two heart fields touching.) The field aligns itself roughly along the spine, from the pelvic floor to the top of the skull. It is much like the Earth's magnetic field, which is aligned from the North Pole to the South Pole.

Heart cells entrain not only with each other, but the heart (and its field) can entrain with any other EM field it encounters. At that moment of entrainment, when the two fields begin oscillating in unison, there is an extremely rapid exchange of information. As the information from each heart field is taken in by the other, heart function alters, hormonal cascades change, and alterations in physiology occur. In essence, a kind of dialogue occurs.

This dialogue is extremely natural to us because it is one of our earliest experiences of life. In our mothers' wombs we are immersed in the field of the mother's heart. The emerging heart of the infant entrains with mother's and continues to do so after birth as well, during breast feeding. The mothers EM field is filled with information: how she feels about the infant, whether it is loved or not, wanted or not. The mother's meaning-filled feelings alter the shape of her EM field and the baby takes the information in and decodes it, much as a radio receiver takes in and decodes a pattern of radio waves.

We are always sensitive to EM fields after birth for we gestate in the midst of this kind of language. After birth, the heart routinely scans encountered EM fields for information. We experience these fields in a unique way; they are experienced as emotions.

Just as the primary colors combine to make up all the colors we see, and basic tastes combine to make up all the tastes we taste, basic feelings—mad, sad, glad, scared—combine to make up all the feelings we experience. The particular spectrum of EM energy that the heart takes in is not experienced as colors or sounds but as emotions. The slightest change in heart function creates new feeling complexes while the slightest change in emotional state creates new heart rhythms. Both immediately show up in electrocardiograph (ECG) and magnetocardiograph (MCG) readings. The heart is, in fact, an extremely sensitive sensory organ whose domain is feeling. The nuances of emotion that it is possible to experience from encountered EM fields are as diverse as the range of colors or tastes we experience. Unfortunately, this finely tuned emotional perception of the world atrophies in those of us who habituate consciousness to the brain. (One of the simplest ways to initiate thinking with the heart, and to begin to reclaim this kind of sensing, is to look at something that is in front of you, a plant perhaps, and ask yourself, "How does it feel?" You will then experience a unique feeling complex, generally not nameable, as the EM signature of the object moves through your heart.)

Living organisms possess extremely complex electromagnetic fields. Each field encodes *everything* about the organism that produces it, its state of health, history, potentials, and much much more. To take a very simple example, every chemical that a plant makes can be identified by its own unique electromagnetic signature. And most plants make hundreds to thousands of different chemicals each and every day. Developing sophistication with heart perception does in fact make it possible to accurately determine the medicinal actions of a plant by directly encountering the plant's EM field. Once we allow the EM field to pass through the heart, it is routed onto the brain for analysis where the meanings in the EM signature are extracted.

Between 60 and 65 percent of heart cells are neural cells, exactly like those in the brain. And the heart's neural cells function in the same way brain cells do; they cluster in ganglia and connect to the neural network of the body through axon-dendrites. This is not an accident; the heart has direct connections to specific centers of the brain and these connections cannot be turned off. There is always a direct, unmediated flow of information from the heart to the brain. The heart is hardwired into the amygdala, thalamus, hippocampus, and cortex. These brain centers are involved with: 1) emotional memories and processing; 2) sensory experience; 3) memory, spatial relationships, and the extraction of meaning from environmental sensory input; and 4) problem solving, reasoning, and learning. To enhance communication with the brain and central nervous system, the heart makes and releases its own neurotransmitters as it needs them.

When the electromagnetic field of the heart entrains with the electromagnetic field of any other organism, whether human, animal, or plant, there is a rapid download of information from one organism to the other. Although this information download is in a language of its own, it is not in words. In one sense, the transfer of information can be thought of as a direct conveyance of meaning without having to use words. This information, as researchers have found, flows through the heart first, and is then routed to the brain through direct heart/brain connections for further processing. In order to retain the information, we convert it into a usable form. A translation process occurs, much like the one our radio receivers provide when they convert radio waves into music. However, in human beings the process is much more complex. From its store of sensory data, memories, experiences, and knowledge, the brain constructs a gestalt of the information flow. Thus the translation can appear in many forms: a series of visions, sounds, images, feelings, tastes, words, or smells. Often the form in which the translation emerges is shaped by the culture in which the individual person was raised. The important thing is, again, not the form that these translations take, but the meanings within them.

These experiences can be intentionally initiated, as the great German poet and botanist Goethe showed. Or, as is much more common in indigenous cultures, they can happen spontaneously. Some examples of these spontaneous events are described in this book—a blending of the EM fields between a human and a plant occurs of its own accord, and in that moment of synchronicity, a direct flow of depth information passes between the two organisms.

The ancient Greeks called this heart exchange with other living organisms *aisthesis*. It literally means "to breathe in." The Greeks recognized that this moment of entrainment between two organisms was accompanied by a gasp or deep inspiration as the impact of the meanings coming in were felt. They considered it an exchange of soul essence, when the thing that is more than the sum of the parts, that thing that came into being at the moment of synchronicity, that thing that is our soul, was touched by the soul in something outside the self.

For people who engage in this kind of exchange, one of the major impacts is the continual awareness that we are never alone. We find ourselves companioned by other ensouled phenomena that care enough about us to engage in this kind of sharing. Through this deep exchange, we are not only anchored in the knowledge that the other inhabitants of this Earth are intelligent, we also experience a direct information exchange with them through our hearts, without the kind of reductionism necessary when consciousness is located in the brain.

Thus indigenous peoples could say, in all accuracy, that they were taught by the plants themselves, or that their knowledge came in visions or dreams. And, of course, this exchange was not limited to plants but occurred with all parts of the natural world. The natural empathy that such close connection engenders also causes people to treat the world very differently than they do when they are alienated from nature. Although explanations of this process of direct perception help us to understand the process, what is most important is the experience of it—how it feels, how it enriches our lives, and how it reconnects us to the ground of being from which we have come. It is very hard to cut down a forest when we experience it as alive and intelligent and as an elder to the human.

In the end, what becomes clear, when we reclaim the heart as an organ of perception and cognition, is that we have been colonized by a particular kind of thinking. And this particular kind of thinking, naturally engendered when we locate consciousness in the brain, reduces our expression of perception and thought. Like monotheism, it reduces what can be experienced to a narrow band of the permissible. All other perceptions are relegated to the superstitious, or heretical, or unacceptable. But this heart-centered way of perception is the oldest we know, intimately bound up in our humanness and our expression as ecological extensions of this Earth. Reductionist and monotheistic approaches can be likened to concrete sidewalks. They suppress the wild, but the power of the green—*veriditas* as Hildegard of Bingen called it—will always break through the sidewalks, both those outside of us, and those within us.

This book was the beginning of my understanding that breaking through, the first articulation of my own experiences with some wild green part of myself reaching upward to the light. In the twelve years since I wrote it, I have traveled much farther along this path. Still, upon rereading it, I have little to argue with. I hope that you can find in it, as I still do, the traces of those who have gone before us, those that called us back to this green world from which we have sprung. I wish you well on your journey.

Stephen Buhner
November 2005

PREFACE
TO THE FIRST EDITION

During the years from 1985 to 1995, I was fortunate to develop a deep, transcendent relationship with thirty-five acres of land outside Boulder, Colorado. What happened to me there motivated me to begin the work that led to the writing of this book. Regrettably, it became clear as time went on that I was not destined to remain on that remarkable land, which had come to mean so much to me. The steady encroachment of people, the micromanagement of people's affairs that the government of Boulder began initiating, and other demands of my chosen spiritual path led to my departure from Boulder. In late 1995 my family and I relocated to Vision Mountain, Sun Bear's former home, outside Spokane, Washington. But the lessons learned during that decade remain strong, as I suspect they will throughout my life.

Many of the things I learned resulted in my coming to use certain terms in specific ways. The word *Earth* is capitalized because, within my own framework and the framework of sacred plant medicine, the Earth is a living being like you and me and as such its name should reflect this. Plants are often referred to as "relations." That is because within this framework all things are considered to be alive—in fact, to be relatives of human beings. To speak otherwise, to relegate them, for the sake of linguistic convention, to a place within the hierarchy commonly accepted in the world that places humans at the apex, would dishonor them.

The spelling of some tribes and societies changes from place to place in the text. The sources I reference often used widely varying spellings; when I talk about a particular tribe or society, I use the spelling of the source being quoted.

I tend to use "holy person" or "medicine person" interchangeably within the text. Some readers may disagree with this usage as indigenous cultures were sometimes specific about the two not being interchangeable. In some cultures, a holy person did not engage in healing work (or did not use herbs), while a medicine person did. Within my own framework, however, the two are identical because "medicine" is a specific manifestation of the sacred and is therefore holy by definition.

To understand sacred plant medicine is to understand certain attitudes and perspectives toward the Earth and all things on it. The underlying nature of our relationship with plants is a primary one.

Human beings have always had a strong and deeply interdependent relationship with the plants of the Earth. Plants created the atmosphere (as a result of photosynthesis) that allowed for the evolution of oxygen-breathing beings. In many ways, we are only the byproduct of plants' habitation of Earth.

As humans evolved on Earth, they used plants for everything imaginable—food, weapons, baskets, clothes, shelter, and medicine. The subject of this book addresses humans' use of plants as medicine, but in a special sense. It concerns that dimension of plant and human relationship where it overlaps the sacred, and the territory where plants are an expression of Spirit. The human, through the developed capacity to travel in sacred territory, makes an alliance with plants in order to gather knowledge and develop the ability to heal. Thus this book deals with plants as medicines and plants as sacred beings. Once upon a time, the two were not separable.

I use some of the information gathered by ethnobotanists and others at the turn of the century about indigenous peoples' plant relationships.

I share these things because I believe that human beings should remember how to behave in the world, that knowing the sacred in all things, without grandiosity or denigration, is crucial to human habitation of Earth. I share them because those who recorded their stories were afraid that their knowledge would pass from the world. They believed (as do I) in the importance of what they had learned.

In addressing the sacred, I intend primarily to speak of it in general, that is, without reference to any particular religious map. I do, however, focus mostly on one particular spiritual lineage, that of the Earth-centered way. In describing this spiritual orientation, the term "Earth-centered" has been specifically chosen over other, somewhat similar terms such as nature worship, pantheism, Wiccan, paganism, and Indian religion. Wiccan and Indian religion are specific forms of Earth-centered practice, much as Lutheranism and Catholicism are specific forms of Christianity. My orientation is neither of those.

Pantheism in its primary meaning, "The doctrine that the whole universe is God, or that every part of the universe is a manifestation of God," is probably the most accurate; however, it has been degraded in common usage to its secondary meaning, "Worship of all the gods of various cults," and as such would be misleading.

Nature worship also is inaccurate in that it implies that what is worshiped is nature rather than the sacred through its manifestation in nature. This confusion was reflected in the writings of many nineteenth-century Christians. In denigrating Native American and other religious forms, which had much of their focus in nature, they often argued that the obvious indication of the inferiority of nature-focused religious forms was their confusion between the Maker and His creation.

Paganism as a term is rejected because many Earth-centered practitioners view it as a denigration once used by Christians during the forceful destruction of non-Christian religious forms before the twentieth century. To many Earth-centered peoples, it contains highly negative connotations.

My intent here is not to focus on religious forms that are attached to a specific cultural background, but to try and convey an ancient and long standing form of Earth-centered spiritual practice—sacred plant medicine. I am concerned less with religion (the bureaucracy of a spiritual lineage) than with the spiritual experience that underlies such religion. By focusing much of the book on the Earth-centered spirituality of North America, some aspects of religious practice will be developed; in particular, certain aspects of what sometimes is called Native American religion. Native American religion is, however, only a specific form of Earth-centered practice; there is no generic Native American religion *per se*. The various tribes of the United States developed, over millennia, cultural expressions of Earth-centered practice that are unique to them. Some tribes' cultural expression of this path overlap with others quite readily, some do not.

You will undoubtedly note I take strong exception to the way scientific opinion is now commonly accepted in the world. This has caused me to question many of its uses, and this perspective comes through in some of my writing. To medical science especially, I take exception. The presentation by conventional medical science that herbs are unscientific and the remnant of an earlier, more superstitious age, is a grave misuse of science when so much data, much of it gathered in other countries, conflicts with this view. In their search for knowledge, there is an attitude of superiority that many scientists possess, which, if allowed to be expressed without proper controls, can be dangerous. Further, our culture has come to rely overmuch on scientific experts. Many people no longer reason for themselves when faced with opinions from scientific experts, and it is questionable whether such experts really understand the workings of nature any more than anyone else. The misapplication of technology, based on the expert opinions of the scientific community, without regard for environmental consequences, has caused a great deal of damage to the Earth. I, and many others, are beginning to question whether or not science can or should play as large a role in determining the safe application of technology.

Finally, much of the early repression of the American Indian came from the scientific community's support of religious and Euro-American cultural beliefs that the indigenous tribes of North America were uncivilized and ignorant.

But a more pervasive problem exists in the unquestioned embracing of the scientific model. Though other methods of information gathering are very useful, most scientists feel that methods other than the scientific are not valid. Adherents of other forms of learning are generally castigated by the world scientific community. This trend has caused the abandonment of many approaches to understanding the world that, I believe, we as a species need in order to successfully inhabit the Earth.

Numerous authors have expressed concern in recent years that many writers are romanticizing Native American religious and cultural forms. This is not my intent. Rather, I want only to share a specific point of view that was once common in the world, and to support that point of view with words of the native people who were courageous enough to share them originally. These historical records illustrate a world view that I feel is crucial to our continued habitation of Earth. While I do quote these original sources, I do not speak for any tribe or its cultural lineage of spiritual experience and devotion. I am not qualified to do so. I speak instead as a human being who travels in sacred territory, one who travels the Earth-centered path. I speak of the territory I have found, the Earth-centered devotions that have called me, and of the sacred plant relations.

It is your birthright as well to understand and enter this sacred territory. I invite you to hear what I share here, to go beyond the words, and feel the touch of the sacred territory that lies beyond.

Only to him
who stands
where the barley stands
and listens well,
will it speak and tell,
for his sake
what man is.

—MASANOBU FUKUOKA

1
THE SACRED
AND THE EARTH

All mystics speak the same language for they come from
the same country.

—Saint-Martin

During the 1950s, the time in which I was raised, the extended family was still alive. Most Americans were farmers, and many people lived close to the land on this North American continent, a continent known to many of its native peoples as Turtle Island.

I was raised in Louisville, Kentucky, still a rather small town at that time. I knew four of my great-grandparents and with two of those I was especially close. These two people, Cecil and Mary Harrod, had been born and raised in the latter part of the nineteenth century. They were filled with a spirit and approach to life that is now, in our time, extremely uncommon.

My great-grandfather trained as a physician before the allopaths achieved a monopoly in medicine, before the advent of penicillin and other antibiotics when, among other things, doctors still used herbs in their practice. He was a horse-and-buggy physician in a small town, and he delivered babies at home. He knew the people and they knew him;

his office was in his home. He helped babies into the world and he was there when they left it; he'd been present during much of the life that was sandwiched in between.

By the time I was born he and my great-grandmother lived in Columbus, Indiana. They were mostly retired, though there was still an office in back of the house. They also owned a small farm outside Columbus where the family would gather most summers. To travel to that farm meant entering another world from the one in which I lived most of my life. The farm house was a converted hand-hewn log barn that had been built about a hundred years before. It had been disassembled, moved to the farm, and rebuilt. There were two good-sized ponds, the smaller one in the deep woods.

My great-grandparents came from a different time. That time, connected to the roots of what I think of as the heart of what America could have been, soaked into my bones from the day I was born. It was a time of horses and of making what you have. A time of slow pace and rural life, of being close to the land and being a simple people not hung about with high-tech designer equipment. A time of the grain of wood sloping down the line of a hundred-year-old hand-hewn log. A time of the deep veins on the back of a man's hand. A time of walking in the deep woods and hearing its call—the taste of freshly picked blackberries, the sucking sound a cane pole makes when pulled out of the mud in sudden response to the hungry pull of a fish. A time of finding secret hiding places, and heeding the call of a woman's voice urging haste so the meal will still be hot. The taste of too-sweet iced tea in big, sweating glass jugs. Laughter around the table and tales of times past when these grown elders were but children themselves. And most of all there was the feeling of being loved deep into the soul and not caring a hoot about the shape of a person's body or the irregularities of their personality.

My great-grandparents' way of life soaked into my soul and every fiber of my being. My bones fed on it as did my mannerisms, language, hopes, and dreams. But even more central and enduring was the deep connection with the land that they had shown me. That connection reached into my spirit and told me in its own special language that there was a deeper world, far older, than the human one in which I lived.

But that sense of connection with the Earth was quiet throughout the ensuing years as family died, places were sold, and television replaced the sounds of my elders' voices. As the space of years increased the greater my feelings of alienation became—from myself, from any kind of enduring values, from the Earth. But those early times, the connection to that land, were like seeds in my soul and eventually, without warning, they burst into growth and changed the course of my life.

They were part of the impetus toward my deep experiences of the sacred when I was seventeen, part of that which led to my trying to understand analytically what had happened to me, part of the map I then created. And eventually, as I began this journey into the world of sacred plant medicine, their threads urged me on to a deeper understanding of my relationship with the Earth.

There is a dimension of human experience, a way of experiencing the world, where plants can talk to human beings and humans can talk to plants. It is a dimension that has been utilized by human beings for most of our history on this planet. Yet in these days and times, that way of life has been relegated to a back alley in a poor neighborhood of the city of man. Those who live there are generally thought of as ignorant and uneducated, superstitious and primitive.

The knowledge that human beings found in exploring that dimension of life, in talking to the plants and hearing the plants talk to them, predates the spread of what today we call science. In that dimension of human experience people commonly sought close contact with the sacred, to know the desire of Creator, and to bring into the world the spiritual visions given them by Creator. Knowledge gained in this manner worked well (as scientific study has shown) and looking over at it from this world of science in which we now live, many are amazed that any of the knowledge found in that other world worked at all. We no longer understand that kind of information gathering; in the industrialized nations we have given it up. In many ways we are the poorer for it. Still, in the backwaters of the world people live as they have for thousands of years, in relation with all life and all living things. As Mother Theresa said, softly, when receiving her Nobel Prize, "It is not we who are poor but you."

The poverty Mother Theresa spoke of can be felt strongly when comparing the following stories:

In Korea it is said that in the year 850 A.D. the ginkgo tree was in danger of becoming extinct. It is a tree whose existence is interwoven with that of human beings in Asia, only recently (around 1800) having

Ginkgo tree
in the temple

been introduced into the West. The ginkgo is used for food and medicine and additionally is held to have many spiritual attributes. In that time of danger, many Buddhist monasteries in Korea began taking in saplings to protect the tree from extinction. The Buddhists are credited with saving the ginkgo by taking it into their temple gardens.[1] One of the largest ginkgo trees in Asia grows on the grounds of Yongmun-san temple in Korea. It stands 180 feet tall, 15 feet in diameter, and is said to have been planted in the ninth century. This tree, planted to protect the species from extinction, was kept with reverence and prayer for over a millennium. People still make pilgrimage to visit it.

Some 1100 years later (around the 1950s), a graduate student was finishing his Ph.D. on the bristlecone pine. He had completed his course work and was working on his dissertation, conducting field research in a bristlecone pine forest. He was trying to establish the bristlecone as one of the oldest trees on the North American continent.

He hiked for many days, packing in his equipment, and set up camp. Eventually he located the tree he would study, the tree that he believed was the oldest tree in the forest. In fact, he intended to use a core drill to extract a sample from the tree in order to count its rings and establish its age. He kept extensive field notes and made careful preparations. His Ph.D. degree depended on the research he was conducting on this tree. However, when he was ready to drill a core from the tree, the core drill was not working. He struggled with it for several days trying to fix it but to no avail. He did not have another core drill but he did have a saw. And he cut down the tree and found it was some 4000 years old by counting its rings. A tree already 2000 years old when Christ lived, cut down for a Ph.D.[2]

It is never possible to share this story without experiencing deep feelings of sorrow. The story brings home the rift between humans and the world, the poverty and illness with which we struggle as a species. The feelings it brings to the heart are often too deep for words. Yet, they are ones all of us try, often with great success, to repress. There is not a one of us who does not know there is something amiss in our world. There is not a one of us who does not carry the grief concomitant with the damage to our home. As Aldo Leopold notes:

> One of the penalties of an ecological education is that one lives alone in a world of wounds. Much of the damage inflicted on the land is invisible to laymen. An ecologist must either harden his shell and make believe the consequences of science are none of his business, or he must be the doctor who sees the marks of death in a community that believes itself well and does not want to be told otherwise.[3]

A solution to the poverty and illness in our world lies within the ancient capacity for individuals to travel in sacred territory, to reconnect with the sacredness of the Earth, and to develop their own capacity, a birthright of being human, to evoke the holy and once again sit in the council of all life.

The Sacred

Our capacity to recognize and seek out the sacred is one of the basic drives that make up the fabric of a human being, thus shaping our common human ancestry. The sacred as I use it is more akin to the dictionary definition of *holy,* "having divine nature or origin." It must be recognized that because the sacred is made up of both nonrational and nonlinear elements, any reduction to simple definitions always fails to capture its essence. One must enter the realm of the sacred and experience its transcendent nature to fully understand it. There is a distinct reality that underlies all religious articulations. It is this reality that, when experienced, is felt to be the REAL, a deeper and more meaningful reality than that we experience in our normal day-to-day lives.

The maps that travelers create from their travels in sacred domains, and the bureaucracy that springs up around control over the map, make up the form and substance of religious movements. The maps correspond to specific lineages of religious or spiritual devotion. All humans have a propensity for *how* they experience the sacred. For example, human beings may experience the sacred as a territory (Native Americans), as a personification (Christians), or a state of mind (Buddhists). This propensity for how one experiences the sacred can lead to arguments (and sometimes wars) over the correct way to experience the sacred, over "The Way." But as the eminent religious historian, Mircea Eliade, has said, "There are no definitional limits to what forms the sacred can take."[4] The manifestation of the sacred—hierophany—may occur in any person, place, or thing. The sacred, by definition, can take any form.

Each religious articulation has its place within the human frame. To claim superiority for a religious expression is to claim the thumb superior to the fingers, the foot superior to the leg. Each has its necessary place and function. One must search for the real center of religion and go beyond the linguistic representations contained in religious maps. If one does not, one finds the human, not the sacred.

The sacred has a dynamic aspect in that it has a tendency to manifest itself of its own accord. It tends to come into the world and make itself known. Further, each incarnate form, each object of matter, has a tendency to realize its archetypal, universal, sacred meaning. These two tendencies—that of the sacred to manifest itself and that of each incarnate form to realize its deeper archetype—come together in such a way that any object at any time can incorporate within itself all the power of the holy. When the sacred manifests itself in the world, something in the human allows it to be immediately recognized. A part of the human, most often a subconscious part, experiences the sacred and says to the conscious mind, "that is the REAL." The conscious mind is then made aware of that which is beyond it and that from which it comes, the sacred.

The intrusion of the sacred into human experience represents a direct transmission of the REAL, a transmission of God, Creator, Allah, Great Spirit. The human who experiences this is made aware of a reality that transcends the human and thus predates human linguistic and cultural constructs. This presents difficulties. How does one retain the memory and experience of something that predates all things human? To explain the experience and to retain memory of it, human beings automatically structure the direct experience of the sacred into internalized symbolic constructs. Thus the sacred comes to be expressed in visions, wondrous feelings, thoughts, and sometimes smells and tastes. This is due to the nature of memory patterning.

Human memory patterns are constructed from aspects of the five senses; that is, memories are encoded bits of sights, sounds, smells, tastes, feelings. Thus the experience of the sacred is translated into visions, sounds, smells, tastes, and feelings even though the sacred is both all and none of these things. Examinations of the written and oral records of those who encountered the sacred show that their experiences were very rich and generally included all of the five senses.[5]

Strong visionary experience is often accompanied by imperatives for human conduct. Conveyed during contact with the sacred, these imperatives often require the person to whom they are given to act in a certain manner, engage in a specific life work, or make changes in lifestyle or behavior. Because these imperatives are usually interpreted as language when experienced, they most often take on the pattern of language that is already encoded in the person receiving them. To make the imperatives sensible, people also interpret them through previously learned cultural experiences and values. Thus, if one is raised in a primarily Christian environment, any direct experience of the divine will often tend to take on Christian forms and symbols.

All these things—sensory memory bits, linguistic and cultural structures that give the experience of the sacred memory form—become symbols that contain in themselves the capacity to reinvoke the original sacred experience. Though these elements are used, the sacred does not become only those things. Inherent in the experience of the sacred is the memory of its transcendent nature, and according to their capacity, humans are forced to generate more powerful constructs out of their own existing structures to encompass the immense morphology of the sacred. In this process it is not possible for the human to retain the full experience of the sacred. It is too large a territory. Even so, the human has been changed, is no longer only secular, and the symbols retained point the way to something other and more REAL than the human.

Within many cultures, the search for personal contact with the sacred is an integral part of our maturation and development. When contact with the sacred occurs, its nature and content shape the direction of that person's life. It provides meaning by which that person determines ethical and honorable behavior and life's work. Further, frequent contact with the sacred through personal visionary experience or community ritual gives direction for the deepening of one's own spirituality over time.

Though experiences of the sacred cover a wide spectrum of styles, the oldest and most widespread is Earth centered, or what is sometimes referred to as pagan religion or nature mysticism.

Earth-Centered Spirituality

To Earth-centered peoples the sacred is immediate. It is present in all parts of the world and one may, simply by being willing to be in relationship with the deeper aspects of a part of the Earth, attain closer relationship with Spirit. Through this closer relationship can come knowledge that gives guidance and meaning to one's life and community. Through this deeper relationship over time one can gain power to evoke the sacred through ceremony, to shape its course into human affairs to benefit the community, to heal and instruct, to uplift.

For those on the Earth-centered spiritual path, the Earth itself is *the* place of worship; all things possess a soul, every tree, stone, and root. To those in relationship with Earth, the Earth *and each part of the Earth* is a manifestation of the sacred, a creation of Spirit. Both the whole of Earth and each part of it can manifest itself as the *ganz andere*, the totally other.[6] Within Earth-centered cultures the stone and the tree are venerated as an expression of the sacred, a creation of

God. Oftentimes, either through ritual or through direct manifestation of the sacred, the stone or tree begins to exhibit a totally other reality. It begins to manifest the sacred archetype of its kind. Far from being stone or tree or plant, it becomes STONE or TREE or PLANT, a pipeline to a deeper and more meaningful reality. And within Earth-centered religious practice, individuals spend many years, sometimes a lifetime, in developing deep relationship with the aspect of the sacred that has revealed itself.

Within the lineage of Earth-centered spirituality, the individual will never choose which aspect of the sacred to focus on. In some manner, the sacred world indicates that the individual should focus on this or that aspect. This may come through visions (the sacred intruding itself into the individual), through hierophany of place (the sacred intruding itself into secular space), or through "signs," i.e., indications and tendencies in self, family, or environment that direct one to a certain path.

For someone who is touched with the underlying reality of BEAR, the focus of spiritual life and power and communion with Spirit comes through the deeper archetype of the bear. The sacred and the archetype of the bear speak and convey knowledge and power to the person who has such deep contact with BEAR. Over time one is able to use this personal connection and knowledge of BEAR to manifest the archetype of BEAR into normal, secular reality and thereby effect change. Such a person is one of spiritual power but this power is specific in its scope and use. The power may be usable only when certain illnesses are present or certain ceremonies are necessary or the community faces certain crises. In other times and needs, the power of a specific archetype is not accessible or effective.

Human beings have a deep affinity for the sacred and all its manifestations. This is especially true of Earth-centered peoples. The process of the sacred manifesting itself in an incarnate form, the tendency of the form to take on its sacred archetype, is well known to Earth-centered peoples. To them, life is a love affair with the constant expression of the sacred in all its forms. To be in close proximity with the sacred is to be close to the REAL, to become saturated with the power of the holy. In the long run, this allows one to become REAL, to embody or carry the power of the holy in oneself.

Ecological Influences of Earth-Centered Spirituality

Relating to the natural world in this manner tends to create ecologically sustaining behavior. It is very difficult to hunt to extinction a species that is felt to be a relative or a sacred expression of Spirit. In industrialized societies, many people have forgotten this way of feeling and the Earth and its parts have become only resources or objects of study. This depreciation of the value of life has begun to have grave effects on our Earth and on our lives. In denigrating the value of other members of the ecosystem we have devalued ourselves. Only our capacity to revere and feel the connection of life can restore the Earth and our own lives.

In Asia there has long been a recognition that certain places on the Earth are more auspicious than others. To Koreans, for example, it is important to build cities and homes and to place graves in such a manner as to ensure harmony. This recognition has been codified and developed into the science and art of geomancy. Geomantically auspicious places are usually mountainous, have good, strong, old forests, and an abundance of plant life. Because of this strong recognition of the life force of the Earth and the desire to maintain auspicious places, geomancy has become a strong ecological force in both Korea and China. Industrialization and the assimilation of Western attitudes and lifestyles have eroded the prevalence of geomancy to some extent, yet it still remains a part of Asian life.

Hong-key Yoon, in his seminal study on geomancy, *Geomantic Relationships Between Culture and Nature in Korea*,[7] reviewed Korean government records of 2000 years duration for decisions affecting the ecology of the region. He found that periodically the Korean people would overharvest a particular region, thus affecting its geomantic auspiciousness. Decisions were made declaring such regions off limits for harvesting for 150 years or so to allow the tree population to regenerate and restore its geomantic harmony. Yoon was able to track this cycle over and over again in Korean history. This sense of the sacredness and power of the Earth periodically ensured that specific areas of Korea would be allowed to lie fallow and free from human intervention for more than a century. Though laws were involved in this process, the laws themselves only reinforced the underlying beliefs held by most Koreans. This act of setting a place aside and holding it in reverence is not new, nor is it particular to Korea.

Similarly, the cosmology of many preindustrial cultures makes regular allowance for the experience of sacred power or hierophany in the Earth or elements of the Earth ecosystem. This sense of the sacred manifesting itself in certain areas has routinely led to preindustrial tribes setting certain areas apart as sacred or taboo to human intervention.[8] These areas remain undisturbed and provide places for those feeling the need to seek closer contact with the sacred. The result of this, ecologically, is the uninterrupted existence of large areas of wilderness that contain exceptional degrees of diversity.

World Watch researchers noted:

> South Asian and Southeast Asian farmers have traditionally honored sacred groves, believed to be homes of powerful deities. The Kuna and Embera-Choco Indians of Panama leave patches of old-growth forest as supernatural parks, refuges for both wildlife and spirits. Waterways as well as forests are protected by the Tukano Indians of Brazil, whose taboos guard as much as 62 percent of nearby streams as fish sanctuaries.[9]

A 1989 Sierra Club survey of large wilderness areas found that over one-third of the Earth's land surface, an area in excess of 4.8 billion hectares, is still controlled by natural forces, not human. The survey noted that the healthiest ecosystems still remaining were all in these areas and that these areas and indigenous territories were very nearly the same.[10]

There is a basic interrelationship between the health of the Earth and the experience of the sacred. As connection with the sacred grows more tenuous, fewer and fewer areas of the Earth will remain alive and wild and sacred.

This desacralizing of the world is then a modern phenomenon. Only the United States and a few other countries have no integral mystical tradition that is given cultural recognition and support. The human tendency to experience the sacred is not lost but in such cases, it becomes a debased and furtive experience. Those people who have mystical experience have no cultural guidelines to direct their development, no elders. They cannot overtly be recognized for their condition (being in contact with the sacred) nor can they bring the knowledge and renewal of life energy they gain from such direct spiritual experience to their community—one of its main purposes. As a result, the culture itself begins to lose a sense of meaning and direction and the land on which the culture is founded becomes increasingly degraded. For a society to remain viable, its members must frequently renew their connection with the REAL.

In renewing connection to the REAL one directly encounters the nonrational elements of the sacred. One travels into a realm where common boundaries and definitions are nonexistent. One seeks a different order of *meaning*. One finds it in common everyday objects and events set apart only because they now express the sacred. Thus one lives in a world that is simultaneously sacred and secular.

To conceive of a territory overlaid with another territory of the same size and shape but that has a different order of *meaning is* difficult for modern sensibilities. But in many cultures, and all Earth-centered ones, it is common for human beings to travel into sacred territory to bring back knowledge of the meanings they find there. Part of that knowledge shows that certain parts of the Earth are sacred and should be set aside and revered.

When human beings travel in sacred domains they not only have the inherent capacity to recognize the sacred but also to recognize patterns of sacred expression. Sacred travelers often focus on a particular aspect of the sacred world, a specific pattern of sacred expression, and spend their lives in devotion to understanding Spirit through that particular aspect or pattern. One example of this is the expression of the sacred through plants. The sacred expresses itself through plants in ancient and well defined ways. To Earth-centered peoples this has always represented one manner of becoming a holy person, one way to find knowledge of God, one way to bring back the immense power of the holy to one's community.

The names of most of these holy people will never be known. But those ancestors have laid down, in the human DNA and in the landscape of the sacred, pathways one can find and travel. The formidable aspect of traveling in sacred territory is not diminished but the work is made easier by knowing that our ancestors have come this way before; it is possible to find the traces of their passing. As Crashing Thunder's father said:

> If you wish to obtain real blessings, so that you can cure even more people, you will have to fast a long time and sincerely for these blessings. If four or perhaps ten of the powerful spirits bless you, then someday when you have children and anything happens to one of them, you will not have to go and look for a medicine man, but all you will have to do will be to look into your own medicine bundle. Look therein and you will be able, with the medicine you find, to cure your children of whatever ailments they have. Not only that, but after a while you will be called to treat your fellow men. Then you can open your medicine bundle and not be embarrassed, for you will know how to treat an individual who is ill and needs medicines since you will possess those that are good for him. You will know where the seat of his trouble exists, and since you will have obtained these blessings only after the greatest effort on your part, whatever you say and do will be efficacious. If you declare that he will live, then he will live. If you make proper offerings to

your medicine, and if you speak of your medicine in the way you are accustomed to do, and if you ask your medicine to put forth its strength in your patient's behalf, the medicine will do it for you. If, in truth, you make good offerings of tobacco to your plants, if you give many feasts in their honor, and if you then ask your medicines to put forth their strength, and if, in addition, you talk to them like human beings, then most certainly will these plants do for you what you ask. You can then accept the offerings patients make to you without any embarrassment and your children will wear these offerings and will gain strength from them. They will be well and happy. So be extremely diligent in the care you take of your medicines. Medicines are good for all purposes. That is why they were given to us. We are to use them to cure ourselves. Earthmaker gave them to us for that purpose.[11]

2
TO BE HEALED
BY THE EARTH

I awoke one morning in early 1988, doubled over in extreme pain from intense abdominal cramping. For about 20 minutes or so I lay there unable to move, wondering what was happening to me. Then the pain subsided. I was exhausted and covered with sweat. Once I had recovered I imagined the cause to be anything from food poisoning to appendicitis. But once it had passed I seemed much better and, as there was no immediate recurrence, I thought no more of it.

About a month later, I had a similar attack of cramping. Again, my concern subsided soon after the pain was gone. However, when a third attack occurred near my physician's clinic, I gritted my teeth and drove quickly to his office. He saw I was in pain and agreed to see me immediately. After poking and prodding me for awhile, he admitted he didn't know what was wrong, charged me $45, and recommended that I see a specialist.

The specialist, a surgeon, arranged to squeeze me into his busy schedule—two weeks later. On the day of our appointment he had me lie face up on an examining table. He began by pushing my abdomen down toward my backbone and then letting it go with a sudden jerk. The pain was incredible. When I asked what he was doing, he explained that he was checking for "rebound pain." Over the next fifteen minutes, the specialist and several enthusiastic assistants subjected me to a series of excruciating tests. Systematically, they ruled out appendicitis, ulcer, gallstones, and kidney stones but they couldn't determine what was wrong with me. With barely concealed enthusiasm they suggested I consider exploratory surgery. I agreed to think about it, paid their $150 fee, and left. I didn't know what I was going to do.

Coincidentally, I had recently met a local herbalist who had shown me a plant called osha. She told me that the plant was often used for abdominal cramping. The plant grew in abundance around my former home near Boulder, Colorado, and inspired by the synchronicity of the discovery, I decided to try osha. I wasn't going to opt for surgery.

I dug up and dried several roots and began carrying one with me, nibbling on it throughout the day. When the next abdominal attack came I took a fairly large quantity of the powdered root in capsule form. The pain was noticeably less intense and quickly subsided, so I decided to continue taking osha. Over the next two months, the attacks lessened progressively in intensity until they finally stopped entirely.

As soon as I started taking osha, I sensed that this wild medicine was qualitatively different from any pharmaceutical drugs I had used in the past. As I continued eating a little of it each day, I began to feel the spirit of this wild plant working inside me. I knew intuitively that osha was a living being, and that it had an intelligence of its own. Most important, I could feel its spirit interacting with my body in such a way that I knew it was healing the disease inside me. With this came an incredibly strong revelation: for the first time in my life, I experienced the Earth healing me. I was having an experience of the Earth as a living, conscious being; the concept was no longer an intellectual one. I had made a transition from the abstract to a living experience.

Being healed by a plant that grew wild in the fields and valleys around my home was a transformative experience. Although I had been involved in various aspects of Earth-centered spirituality—as teacher and as student—for nearly 20 years, I still viewed wild plants in much the same way most Americans do. I knew the names of some wild flowers, and I could identify a few wild plants that produced edible nuts or fruits. Apart from that, I tended to lump wild plants into three categories: big ones (trees), medium-sized ones (bushes), and little ones (grasses and weeds).

Now I wanted to learn all I could about the world of plants and plant medicines, and especially about their underlying sacredness and spirit world. For the next six years, I spent much of my time exploring the remarkable world of plants around my home. At first, I focused on studying the plants' physical properties. I spent several hours each day identifying and observing the wild plants growing in the nearby fields and woods. Gradually I came to know a great many of the local plants—in all seasons and all stages of growth. I came to know them intimately. I learned their environment, how they like to grow, their preferred terrain. I felt their roots with my hands, tasted them with my mouth, smelled them with my nose, experienced their healing in my body, and used them to treat my family's ills. They came to permeate my life.

My deep experiences with osha had remained strong for me. I wanted to see if I could replicate them with other plants. I began spending several hours a day deepening my relationship with different plants. Sometimes I sat quietly with a plant. Sometimes I kept an image and feeling of the plant in my mind as I carried out other tasks in my daily schedule.

As I spent weeks, and often months, with each plant, I came to be able to distinguish one "feeling tone" from another. I worked with this for over a year until I could distill the essence of each feeling tone into usable knowledge. That is, I learned what essential elements lay at the heart of the feeling tone, those things that gave it its distinct emotional flavor, different from all others.

Each species and each plant within a species possesses a distinct energy or life essence that I experience internally as a feeling. These feelings are much subtler than the more readily identifiable emotions of anger, grief, joy, and fear. They represent a wide spectrum of emotional shadings, each distinctly representative of a particular plant or species.

For example, distinguishing such feeling tones is easy to experience internally if you think of taking ten people at random from a mall and having them line up before you. Without thinking, you look at each one and describe the overall feeling you have when first perceiving them. This feeling tone is often distinctly different among people of similar backgrounds, dress, and socioeconomic class. People, often quite unconsciously, use this facility of intuition when "taking to" someone or not. This facility to distinguish specific feeling tones does not begin and end with human beings but can be applied to anything in our world. I had used this technique in my own teaching and healing work for many years but this was the first time I had used it with plants. I found that all plants, as indeed all living things, possess a distinct kind of energy or feeling tone that can be perceived if one directs personal awareness toward sensing it. This sensitivity seems to be an analogous process to that experienced by most people when encountering a puppy. There is an almost instantaneous emotional response that is unique and specific to the human meeting a puppy and, with practice, this process can occur consciously with anything in our world.

Over the years I have cultivated that basic process through internal meditation and experimentation until I can stimulate that reaction into activity. Also, in time, I learned how to focus and direct the feeling tone of the plant into the process of making medicine and during its use by someone as medicine. I learned that if I kept this feeling tone or energy of the plant at the forefront of my consciousness, I could enhance it by increasing my sense of it until it took on, in my own body, the same large proportions of anger, fear, joy, or grief. This capacity seems to be dependent on the directed intention of personal will.

Perhaps one of the most striking examples for me in my exploration of this phenomenon with plants was the first time I encountered angelica. A relative of osha, angelica is somewhat uncommon around Boulder. I rarely encountered it in abundance. It grows along shady streams at the bottom of north-facing slopes. The first time I met angelica, I was walking along such a stream, following a small game trail used mostly by bear and deer. My senses were filled with the environment in which I found myself: the sounds of the stream as it hurried over the stones, the deep greens marked with stripes of bright sun where the light filtered through the canopy of forest, the inexpressible smells of old forest and lush understory. As I turned along the edge of the path, my whole attention was captured by a plant growing next to the stream. It stood nearly six feet tall. A strong, rather large stalk was crowned by large softball-sized clusters of seeds and large splays of leaves standing out from the plant at irregular intervals up the stalk. The plant itself exuded a special quality of its own that was strong enough to make me pause from my exploration. I made my way to the plant and sat down near it. I would say that the major feeling tone that I felt from the plant was a quiet and pervading dignity and maturity of spirit. I spent perhaps an hour there that first time and over the years I have spent many hours with her children—grown from her seeds in that same place—and with their children. I have come to understand the underlying qualities of the plant, those things that go to make up its particular feeling tone; it has become one of my most important plant allies.

Angelica possesses a supreme dignity and sense of wholeness. In some manner it balances the polarities often encountered in nature and the human condition. Its nature is completely and totally female in the deepest sense of the word.

After I had been sitting with the plants for several years, I often found myself entering a kind of waking-dreaming state where the plants began to speak with me about their uses as medicine. One of my most profound experiences occurred with a plant called usnea, a widespread lichen that grows on trees throughout North America. Usnea is strongly antibiotic, to some extent rivaling penicillin. As I was sitting with the plant, focusing on its feeling tone, I entered a state of mind similar to just waking or drifting off to sleep. My vision was softened, colors seemed more enhanced. The feeling tone of usnea, usually subtle, increased in intensity of its own accord until I felt myself lost in it, my normal sense of personal boundaries dissolving. At that moment the plant appeared to me as a youngish man, hair curled and growing like the plant itself. He smiled and told me that usnea's primary function in the Earth ecosystem is to heal the trees; that it acts as an antibiotic for the lung system of the planet; and that its effects in humans are only a byproduct of its intended effects for the trees. He went on to tell me that usnea is specific for infections in any lung system. He left and I gradually became aware of my surroundings again, waking from the state I had entered.

Earlier in my life, when I was seventeen, I had experienced a strong series of visionary episodes, which, at the time, were quite unsettling. These first contacts with deep sacred experiences showed me the fabric of another manner of living. They caused me to embark for many years on a quest for an understanding of what had happened. This became the impetus for my undergraduate work in cross-cultural religious traditions years later. As the years pass such experiences continue to occur. I rarely find them unsettling but these experiences with plants began causing special problems of their own.

Influenced by a world view that tended to see the various elements of our planet as commodities for human purposes alone, I was becoming acutely aware that wild plants were not only living beings but that they had individual destinies separate and unconnected to humans'. The more plants I connected with, the more I found myself in a world peopled by a multitude of living beings, the existence of which I had, for most of my life, been unaware. Further, the plants were, to a great extent, unconcerned with human desires and needs.

But still, I loved this new world that was opening up before me and I asked the plants that they begin teaching me more deeply in this manner. They were hesitant in their answer, finally agreeing, but requiring certain conditions from me before we could proceed. If I wanted to be able to talk with plants and hear their responses, I must agree to treat them with respect and as equals; to speak out on behalf of plants whenever I taught; to help their voices be heard again in the human world, as they once had commonly been heard when indigenous peoples populated much of this continent. These agreements were easy to make, but they have been sometimes difficult to carry out. As I later discovered when examining indigenous traditions, such stringent conditions are an integral part of sacred plant medicine the world over.

Still, a problem faced me. I had to come to terms with the idea of talking with plants. The idea that knowledge can be gained through sacred experience is alien to our contemporary world. In having come to terms with my own deep connection with sacred states of mind, I was keenly aware that receiving knowledge through direct spiritual experience is highly unusual in our culture—so much so that it is considered abnormal and sometimes even pathological.

My experiences with plant medicines were creating a double-bind for me. On the one hand, I knew I had been healed by the spirit of osha and I was delighted to be speaking more deeply with the plant spirits. On the other hand, the idea of using plant medicines on the basis of information received from plants themselves challenged even my paradigms of healing. Long an advocate of holistic healing, what was happening to me transcended holism and challenged me to see that information could come from sources other than rational thought.

Like many Americans, I was raised in a strongly analytical tradition that tends to devalue things of the spirit and feelings of any sort. From a spiritual perspective, modern science tends to focus on the surface of things, a single dimension of reality. Because our culture and educational processes are strongly based on science, many of the deeper elements of human and sacred experience have been culturally lost to a view that cannot transcend this single perspective of reality. The sacred, which contains both nonlinear and nonrational elements, is often in conflict with perspectives dependent on statistical analysis. Yet, like many others, I feel the pull of both these worlds and a desire to reconcile them if at all possible. Because of this internal conflict, I longed for an analytical model of what I had been experiencing, and to find a sense of my place within an historical framework of human experience.

My undergraduate degree supervisor, Professor Edward Canda, encouraged me to formulate an analytical model of how visionary mystical experiences happen. My thesis work was, in fact, the development of this analytical model, one that I have used in my teaching for many years now. Now, as my experiential studies with sacred plant medicine continued to deepen, I decided to try and repeat that type of research so that I could supply myself with a model to understand what was happening to me. I found that I was in need of a conceptual map of this new territory of experiencing the sacred. The only place I could find traces of similar experiences was in ethnologists' accounts of the use of plants by indigenous cultures of the Americas. As a result I focused on this area in my research.

When I began my researches on sacred plant medicine, I expected to go to the public library and consult a great number of books on sacred traditions in working with plant medicines. I was sorely disappointed. Most ethnobotanical guides to medicinal plants tended to gloss over or ignore the spiritual dimension of healing with plants. A few scholarly books described the sacred use of hallucinogenic plants among indigenous tribes, notably Weston La Bane's *The Peyote Cult*. But most books on the spiritual properties of herbs treated the subject in a rather superficial, amateurish manner. I was forced to dig deeper.

Occasionally, I came across random references to obscure sources. These indicated that many shamanic cultures have developed deep sacred relationships with plant spirits. Such references were few and far between; it took about five years before I felt I had sufficient sources to begin study in earnest. Although my research eventually led me into a fascinating realm of the deeper spiritual realities of sacred plant medicine, my initial research took me on a journey through a jungle of biased reports and conflicting cultural viewpoints.

Finding reliable, substantive information on indigenous medicine traditions turned out to be extremely difficult. Because many early ethnographers from the nineteenth and early twentieth centuries tended to view Native American medicine beliefs and practices as either un-Christian or unscientific superstitions, their observations on the spiritual use of plant medicines were often superficial and antagonistic. This attitude in the researchers' work seemed to reflect my own internal conflict between nonrational, nonlinear experiences of the sacred and linear, analytical, rational thinking. As a result, my effort to create a map of this new territory forced me to wade through my own Euro-American culture's conflict with the same issue. My research was further complicated by the fact that ethnological reports often focused more on how beliefs differed from tribe to tribe than on fundamental similarities. However, I did find some early ethnological reports and a number of unpublished doctoral papers that included detailed transcripts of first-hand observations by shamans and other native healers. These records provided invaluable insights and observations into the traditional ways of sacred plant medicine.

The literature I reviewed contained the words of the medicine people themselves. In every instance they were quite clear that the primary source of their knowledge was Creator and sacred visionary states. While many indigenous cultures value and use empirical knowledge about medicines passed down from generation to generation, they typically attribute their most effective healing knowledge to visions or dreams of the plant spirit or other helpers. There is, in fact, another path than the rational to the accumulation of knowledge about our world and our place in it.

3
VISIONS OF SACRED PLANT MEDICINE

What must be emphasized at once is that all these [plant]
hierophanies point to a system of coherent statements, to
a theory of the sacred significance of vegetation.[1]

—MIRCEA ELIADE

The visionary source of plant knowledge was acknowledged in many of the indigenous accounts of native healing techniques. As I read their words, one thing was clear: In the words of each of these medicine people a number of truths lay sparkling. Apparent was the deep caring they held for the children who would be coming into the world, the deep reverence they held for the sacred, and that they were describing a detailed territory of the sacred concerned uniquely with plants.

The sacred reveals itself to people in a variety of manners. Among the holy people of many indigenous tribes, one unique manner of revelation concerned visions in which they received detailed knowledge of plant medicines and how to heal with them. The Europeans who encountered the indigenous tribes of North America found this difficult to understand, in many instances rejecting what they were told out of hand. Often the Europeans denigrated what they were told. But it was hard to overlook entirely the detailed knowledge the Indians had of their plant world. Modern research has substantiated many of the traditional uses of native medicines.[2] As the native peoples were questioned more closely about their plant medicines, they uniformly replied that the source of their knowledge came from visions or dreams or that the plant had told them its medicinal use. This created a consistent problem for researchers. Used to a scientific, rational model of information gathering, the researchers had no ready reference for what they were being told.

Manuel Cordova Rios, a young rubber collector in the Amazon, who was captured by a tribe in South America and trained in the sacred use of plants, made a crucial point to his biographer, R. Bruce Lamb, in *Wizard of the Upper Amazon:*

> You must realize my friend, that the deeper we go into this, both written and spoken words of formal language become less and less adequate as a medium of expression. If I could arrange it we would have a session of visions ourselves and then you would understand. But that would take time. Meanwhile we will continue with indifferent words and inflexible modes of expression.[3]

This problem arises whenever people who travel in that territory of the spirit attempt to describe what they have found there. The territory that they describe, however, when it comes to plant medicines, is similar from culture to culture, continent to continent.

They all note that the human, in making contact with the sacred through plants, must go into the world of plants, not as a human who knows everything, but as a seeker who has come to learn. They all share that plants can talk to humans and that humans can talk to plants. They tell that to talk with the plants or any object requires the accumulation of spiritual power and the ability to walk in sacred territory. Among the Sioux:

The ability to make objects "sacred" thus giving them mysterious power, was said to belong only to men who had the ability to talk with such objects and to understand what they said.[4]

And the Mitlenos in Mexico:

The herbs and flowers also talk to her and she to them, a rapport with the natural world which is not visibly part of the lifeways of other Mitlenos.[5]

And the Zuni:

The Zuni live with their plants—the latter are part of themselves. The initiated can talk with their plants, and the plants can talk with them.[6]

And the Papago:

It was customary to "talk to the plant or tree" when gathering a medicinal substance and also when administering it.[7]

Everywhere there is this common element of being able to converse with plants, to communicate between species, to exchange information. There are common elements that Earth-centered people note to be necessary for this communication to occur. For instance, it is considered insulting to treat plants as objects that can be owned; that if you do, the plants will become offended and fail to respond to the desires and needs of the human community. Many holy people have noted it is important to talk to plants as human beings. It is the ability of the holy people, sometimes called medicine people, to be in contact with the deeper spirit of the plant and to converse with it and treat it with respect that defines the healer as a person of power, a person of sacred plant medicine. Medicine people are expected to use that power for the benefit of the human community, not for personal reasons alone.

To get an idea of the mindset necessary to enter this world where plants can talk to humans and the humans to plants, try the following brief guided imagery. This exercise works best if you are outside in a natural place:

Imagine yourself standing in a purely natural setting, perhaps in a meadow or a forest. Develop your images of the forest beyond, the grasses blowing in the wind, the sounds of birds and of nature all around you. Let that image develop as strongly as you can. When you can feel and hear and see and experience yourself in that environment, relax a minute and let the tensions go out of your body. Let yourself feel as if the Earth is a warm generous hand holding you, and that you no longer need to hold yourself up; she will do it for you. Now let yourself experience an imaginary mindset: there are no cities, no "civilization" as you know it. Imagine that you live in this natural place in which you envision yourself. There is no other place to go. Look out and see the trees and plants and grasses surrounding you. This is all there is.

Now imagine becoming ill. There is no one to turn to. There are no pharmacies, hospitals, no doctors, just yourself in the world. Now notice the plants. They can heal you if you know which ones to take and how to take them. Stay with this image awhile. In this experience you recapture some of how our ancestors felt long ago and how many indigenous people feel now.

If you can successfully envision this for yourself, you can also feel how it is to be dependent on the plants for your healing. For people in such a lifestyle the relationship between themselves and the plants is one of dependency. As a result of this kind of dependency there is also an inherent reverence for the plants themselves. The human must of necessity recognize that his or her survival depends on the plants. This recognition automatically engenders a respect for the plants and a knowledge of the dependency of humans on the environment.

Now try another image: Imagine yourself living in the cities of today. Let yourself take on the mindset that you normally have in the world. All your medicines come from the hospital, doctor, or pharmacy, and are manufactured by humans. If you are sick, you get a pill. Let yourself experience how that feels now that you have something to compare it to.

In this latter perspective, you are dependent on nothing related to the Earth. There is no sense of need here; the human controls all things. This framework is the one in which most people live. It engenders little or no respect for the Earth because there is no inherent need of the human for anything else except for things made by other humans. When people in this mindset look at plants they see them only as a potential commodity. Plants and their essential components are examined, trials performed, and possible medicinal uses found. It is a strict cause-and-effect relationship—one starts from *A* and goes to *B*.

It is very different, however, when one lives within the first framework or world view. Here, the plants are experienced as more powerful than people. They possess inherent powers and they may use them for the benefit of human beings if approached in the proper manner. Because indigenous peoples lived within this latter perspective, any attempt to understand their relationship with plants is doomed to failure without a deep understanding of how this other world view works.

The ethnobotanists who have studied indigenous uses of plants have often noted that native cultures seem to use plants as medicines with astonishing similarity the world over. But what is really interesting is that cultures all over the world have come to identify certain spiritual qualities in the same plants. For instance, in every culture where cedar is known, it is recognized as having benevolent spiritual qualities and the ability to counteract negative forces. Many ethnobotanists have noted this similarity of use and have called it "simply amazing."[8] When studying plants ethnobotanists can see how an indigenous people could come to use yarrow to stop bleeding. There is a simple cause and effect relationship: apply yarrow, bleeding stops. But it is difficult to explain, using the same reasoning, how so many cultures that have never had contact with each other came to identify the same spiritual attributes in a plant such as cedar. It does not fit into the model that is used by medical science, nor does the manner in which these attributes (and medical

Goose

uses) are arrived at. In every indigenous culture plants are recognized as having unique properties for healing that may be understood through dreams and visions.

People who live close to the Earth and love the Earth and all things that grow upon it believe that there is a soul in all things, including the plants. They believe that the plants, or Creator, or sometimes other powerful spirits can convey direct knowledge and information about how to use plants for healing. The manner of these dreams and visions and their interpretation are incredibly similar from culture to culture around the world. In many cultures becoming a healer and treating the sick was considered improper unless one had a vision or a dream.

Over the last two hundred years, as more and more indigenous cultures have been overrun and displaced, the knowledge they have gathered about the plant world has been lost. In some few cases, fearing this, those with deep knowledge of the sacred began describing their experiences in detail to ethnobotanists and ethnomusicologists such as Frances Densmore. They hoped that their knowledge would be preserved for their children and for all people who loved the holiness of the Earth. We all owe them a debt of gratitude for their foresight and courage.

The following is a beautiful example of a vision that led one such person to become a healer using sacred plant medicine. It was related to Frances Densmore by a Teton Sioux medicine man named Goose in the early 1900s.

When I was a young man I was an excellent marksman with bow and arrows. After coming in contact with the army I was given a rifle and cartridges and never missed my aim. One morning I arose before daybreak to go on a hunting trip. As I went around a butte I saw an antelope, which came toward me and stood still a short distance away from me. The antelope looked at me and then began to graze. I took my rifle and fired several shots with no effect. I fired 16 cartridges and wondered what could be the matter. I put in four more cartridges and fired again, but with no effect whatever. Then the animal stopped grazing and began to move slowly away. Then I heard a voice speaking three times, then a fourth time, and the voice said it was going to sing something, and I must listen. The voice was above me and commanded me to look at the sun. I looked and saw the rising sun had the face of a man and was commanding all the animals and trees and everything in nature to look up. In the air, in front of the sun, was a booth made of boughs. In front of the booth was a very bright object and between this and the booth was a man, painted and wearing an eagle-down feather, while around him flew all kinds of birds. The bright object was a sacred stone, and it was heated red hot. After seeing this I heard another voice telling me to look and receive what would be given me. Something in the form of a bird came down, and where it touched the ground an herb sprang up. This occurred three times. The voice above me said I was to use these three herbs in the cure of the sick. The fourth time the descending object started in the form of a bird, but a human skeleton came to the ground. Then the voice above me told me to observe the structure of the human body. I then saw blood run into the skeleton, and a buffalo horn appeared on the back, between the shoulders, and drew the blood from the skeleton. The voice above me said this was a sign that I would have the power more than any other to cure diseases of the blood. The voice came from the sacred stone and said I must use the buffalo horn in curing diseases of the blood, a practice which I have followed ever since. I do not consider that I dreamed this as one dreams in sleep; it appeared to me when I was early on the chase.[9]

In many Earth-centered cultures it is common to seek such visions in special ceremonies called vision quests. Vision quest is a term referring to a time when a person goes into the wilderness, unencumbered with the trappings of culture, and remains there for a time seeking contact with the sacred. Such retreat is beneficial even if no visions occur but often, through fasting and deep prayer, strong visionary experiences will occur. The purpose of seeking such visions is to reinvigorate and renew personal direction and purpose in life and allow one to bring back from the territory of the sacred new knowledge to benefit one's community.

Many visions, however, are spontaneously generated. They come to a person, like Goose, without being sought, sometimes when awake, sometimes during sleep. Within some mystical traditions this is called the state of grace. Such visions would sometimes occur during times of physical or emotional illness. They would impart to the one in crisis enough knowledge of the sacred that healing could occur. Something like this happened to Mary Baker Eddy and led her to found the Christian Science movement. In the Earth-centered tradition perhaps the most famous visions that occurred in crisis were those of the Sioux holy man, Black Elk, recorded by John Neihardt in his book, *Black Elk Speaks.*

Black Elk's vision was very long and detailed, and in it he was also given herbs with which to help his people.

> We were facing the east and I noticed something queer and found out that it was two men coming from the east and they had wings. On each one's breast was a bright star. . . . They had an herb in their hands and they gave it to me, saying: "Behold this; with this on earth you shall undertake anything and accomplish it." As they presented the herb to me they told me to drop it on earth and when it hit the earth it took root and grew and flowered. You could see a ray of light coming up from the flower, reaching the heavens, and all the creatures of the universe saw this light.[10]

After Black Elk had recovered from his vision he took his friend, One Side, with him to find the herb of his vision, called the Daybreak Star.

> We got on top of a big hill and saw a place and knew this was in the vicinity of the place that I had seen in my vision. We sat down and began to sing some of the heyoka songs on this hill. Pretty soon I began to sing the song I had sung in the first vision. . . . When we finished singing this song, down toward the west I could see magpies, crows, chicken hawks and eagles swarming around a certain place. I looked over to One Side and said: "Friend, right there is the herb." And it seemed as though One Side knew it. So we said: "We will go forth and see."
>
> We went down to the Grass Creek and followed it down a little ways where there was another dry creek and we followed this up. As we neared the place, the birds all disappeared. At this place there were three or four dry gulches together and we saw the herb right there on the side of the bank and I'd never seen one like it in my life. It had two stems and had all the colors of the four quarters. I offered some kinnikinnick to the four sides of the earth. I offered a prayer to the herb saying: "Now we shall go forth (meaning the

herb and I) to the two-leggeds, but only to the weakest ones. Take courage, for there will be happy days among the weak."[11]

Visions of the sacred uses of plants are not confined to the Sioux. They are common all over the world and in all parts of North America.[12] Holy people the world over make distinctions about the body of the plant and its spirit. They learn about the plants in visions or dreams (or under the mentorship of an elder) but they say that it is the sacred *properties* of the plants that heal. The body of the plant helps but is not the main factor in healing.

William C. Sturtevant's doctoral dissertation, *Mikasuki Seminole Medical Beliefs and Practices* (Yale 1955) explores this aspect of sacred plant medicine. One time, in the presence of his primary informant, he referred to a mixture of herbs and water as "medicine" and was corrected. He was informed that those herbs were not medicine but a decoction of herbs. They did not become medicine until the medicine person activated them into medicine through ceremony and (in the specific instance of the Seminole) the use of his breath.

The Seminole call Creator *fisah ki':ko'mihci,* which means "Breath Maker" or "Life Maker." The plants' bodies helped the body of the sick person but did not cure him. It is the power given to medicine people through their training and close relationship with Spirit that is infused into the plants and causes them to awaken into medicine for healing.

Sturtevant goes on to note:

> The most successful doctors have a "living medicine" in their bodies which enables them to doctor more effectively than others. When such a man prepares medicine, this material inside him churns around and gives off a "smell" which enters the patient and aids the cure.[13]

This Mikasuki account is similar to that of a Makah (Nootka) medicine man, Young Doctor. In a dream he was taken to the spirit world where he was given many songs to heal the sick. At one point in the dream:

He saw that when a certain man sang, a snake came out of his mouth and entered the sick person, then it turned into a vapor and emanated from the entire body of the sick person, taking the disease with it.[14]

It is the inherent recognition of this world of Spirit that allows human beings to know they are in the presence of the REAL. Once known it can never be mistaken. Once they begin learning in this world, they also begin the development of the capacity to enter the spirit realm at will and evoke it for teaching and healing. They learn how to evoke the power of the holy to heal.

This capacity to enter the spirit realm is a primary attribute of holy people among indigenous tribes. Once a person begins working with spirit

Young Doctor

powers, the realm of the sacred, and visionary states, it seems to be common that they develop more and more facility for those kinds of experiences. Over time, as they begin learning more and more about the terrain of the sacred, they have the capacity to go further and further into its landscape. This is a crucial element of their role in tribal life. Their forays into sacred territory allow them to bring back information and knowledge to help the tribe and its individual members. Though they might focus on a particular kind of spiritual power, their primary job description might be "One who travels in the spirit realm to gather knowledge for the benefit of his or her community." As has long been recognized among indigenous cultures, the capacity to do this is present in people to varying degrees. The more latent capacity people have to enter sacred territory, the more likely it is that they should pursue this capacity as their vocation.

In our time it has become common for healing to be based on scientific study and trial. Thus healing has become based on information. There is a crucial distinction between analytical study and trial and knowledge gained through sacred experience. To those on the sacred path, knowledge obtained from Creator is much more powerful than that gathered by secular means. To others, it is necessary to legitimize plant medicine or nonordinary healing by reliance on science. Thus sacred plant medicine becomes secularized.

Why did primitive man decide on the use of a plant? It is not so very hard to understand; the plant explained it very well. From where came this capacity of man to understand the plant? From the connection of our common ancestry and our common connection with Spirit. But when people forget that it can be so, they forget an essential aspect of being human. All of us have the capacity to enter this world and know the plants, to hear what Spirit says to them. In each soul, the longing for such experience still remains.

4
THE SACRED SONG OF A PLANT

Shaman song, therefore, represents a profound relation-
ship between spirit and matter. The spirit of breath,
emerging from within the human organism in the form of
song, can be likened to the illumined soul shining through
human eyes.[1]

—JOAN HALIFAX

The closer I have become with the plants the more like people they
seem to me. As I work with them—harvesting them, making
them into medicine, passing them by each day in the fields around my
home—I talk to them, sometimes out loud, often internally. My expression is one of greeting and welcoming, much like the communication
that occurs between two people who care for each other. Over time this
has become almost a talking/humming—a kind of chanting, really—a
song. And with each plant it is slightly different, shaped by the character
of the plant and by my relationship with it.

In studying the history of sacred plant medicine, I found that many cultures believe that all plants have their own songs. It is felt that the power of
Spirit, the plant, or that of healing is experienced and expressed as a song. It

is believed that each plant possesses a song that must be learned; power to heal is given by the conveyance of that song. Among the Seminole:

> Traditionally, Seminole knowledge of curing formulas derives from powerful prophets *(owa':la':ti:)* of ancient times. These men occasionally heard animals or natural phenomena singing, listened, and "picked up" the songs. . . . A prophet might hear a "little bush about 20 yards away," singing in Creek; or during a rain he heard water dripping from the roof and learned the Creek song of the rain; or he heard a fire singing in Mikasuki.[2]

The experience and use of songs were not confined to the Seminole; it was common to tribes further north such as the Cherokee.

> A man once went to the woods and remained in solitary meditation for four days. He wandered alone till he heard a soft, low sweet voice, singing a song. He listened and watched. He saw a beautiful little flower, swaying gracefully back and forth. He knew the song came from the little flower. Around the flower the ground was swept clean. He listened until he had learned the song.[3]

And the Ojibwa:

> It is said that in the old days all the important songs were composed in dreams. A song usually came to a man in his dream; he sang this song in times of danger or necessity. Songs composed in this manner were used on the warpath, in the practice of medicine, and in any serious undertaking of life.[4]

It struck me as phenomenal that this element of sacred plant medicine, that of sacred songs, is so pervasive in North America. In researching the expressions of the sacred in human culture and experience, I found that the concept of the holy expressing itself as music is present in all major religious traditions.[5] But among the indigenous tribes of North America it became part of a long-standing and well-developed tradition.

Perhaps one of the most basic attitudes among North American tribes was the belief that "if a man is to do something beyond human power, he must have more than human strength."[6] To such tribes the concept of a dream took on tremendous importance, and because songs were so often given in such dreams they also represented the power of the holy. As Frances Densmore, the eminent musicologist, has said:

> Dreams and their songs may come to an Indian in natural sleep if his mind is conditioned to such an experience, but the first important dream comes to a young man in his fasting vigil. He is alone in some silent place, and his mind is passive, as he hopes for an impression to come to him from a mysterious source. The silence becomes vibrant, it becomes rhythmic, and a melody comes to his mind. This is his "dream song," his most individual possession. . . . The white musician composes songs addressed to his deity. The Indian waited and listened for the mysterious power pervading all nature to speak to him in song. The Indian realized that he was part of nature—not *akin* to it.
>
> By means of his dream song and by performance of certain acts a man might put himself again in contact with the mysterious powers seen in his dream.[7]

The Importance of Dream Medicine

Within the world of indigenous spirituality, the deep relationship with the sacred powers of the universe through dreaming occupies a crucial and central position. Like Frances Densmore's commentary about song, the Indian also awaited communication from mysterious powers through dreams. These were considered the most powerful of possessions and through them the Indian became a person of power and thus more REAL. Dreams might occur in waking states such as those Goose experienced, or while asleep. But their existence was an active part of every culture's expression. Much of the culture's energy revolved around the possibility of their appearance, their actual appearance, and how they were acted out in the world. Most important cultural ceremonies came from dreams or visions, and nothing of significance in a person's life could be undertaken without dreams or visions playing a crucial part. In the world of plant medicine these dreams were significant. All knowledge of powerful plant medicine came from dreams, though common plant knowledge was exchanged in daily conversation or already possessed by many people in a culture.

Though visions of plants and their uses often came spontaneously to indigenous people many cultures recognized that plants could *make* a person dream. In this instance the plant itself would—by recognizing some unspoken need within people—intrude itself on their awareness and begin calling on them to seek the plant out and learn about it or use it. Jesse Cornplanter remarks on this among the Iroquois when he speaks of water hemlock (muskrat root).

> The old people say that muskrat root is like any other herbal medicine you want. They believe that when you want it, it stands up where it grows calling to you. That is why it is easy to find a medicine which you seek, especially this one with its strong odor.[8]

Although at that time Jesse Cornplanter was being interviewed about suicide among the Iroquois and had been asked specifically about the use of water hemlock for that purpose, he and the other medicine people interviewed noted the connection between an internal need and how the plant world responded. They noted that water hemlock is unlike most plants; other plants "stand willing to respond to man's call for assistance in curing, this one is willing to destroy him."[9] The medicine people related that those who were feeling suicidal would begin hearing the plant call out to them and they would then easily find it when walking in the woods. This is a specific and rather negative action of a plant. In most instances when a person has a need, whether conscious or not, the plant that can help will begin coming in dreams and calling on the person to come to it.

The Sonic Encyclopedia

Many of the dreams that indigenous people experienced contained songs. These songs represented a particular kind of personal power, a holy possession, which were used for as many purposes as the sources from which they came. Songs came from the elements, directly from the plants, and from the animals. Often such animal dreams might concern power in war or hunting or healing.

Because songs were considered to be a source of personal power they were actively sought as possessions. Songs were sought through visions, rigorous apprenticeships with others, or they might be bought, though they were often very expensive. Within some cultures each plant was thought to have a specific song. In such a culture, an apprenticeship in sacred plant medicine might require that the student spend long periods of time with each plant and learn its song. To complete training, the student would be expected to sing the song of each and every plant being used.

Kathleen Harrison remarks on the belief among the Mestizos of Peru in the pervasive nature of individual plant songs and their importance.

> Every species has a song. If you are granted the song in a vision state, or by just submitting yourself to the presence of a plant and opening up, then it's a real gift, and you are able to remember that song forever and share it when it seems appropriate. That song has power, healing power, and there are some which are handed down from one curandero or curandera to the next, and there are others which come to us as individuals. But they are part of an encyclopedia on the sonic level of the same thing that seeds represent on another level.[10]

Harrison's final sentence regarding the equivalency between songs and seeds is a deep and significant insight: ". . . they are part of an encyclopedia on the sonic level. . . ." Dreams and one of their specific manifestations, songs, represent an access point to nonlinear and nonrational aspects of the sacred, a fact long recognized in indigenous cultures. Holy people who specialize in sacred plant medicine—recognizing the access that songs allow to the powerful, nonrational aspects of the sacred—actively seek dreams and dream songs to increase their encyclopedia at the sonic level. The songs in their turn help facilitate entry into the territory of the sacred and their accumulation increases one's capacity to cure, heal, and teach.

Among people who use this encyclopedia at the sonic level there is often the recognition of the power of another's song and at times they might wish to buy that song. If the owner of a song decided to sell it, a price was agreed upon, the owner would "teach the song, explain its use, and show a specimen of the herb to be employed with it."[11]

Each sacred song represents a time when a human being entered the world of the sacred and brought back something from that place for help in the human world. It has been my experience that the most powerful of sacred songs come during particularly potent experiences of the sacred. At such times the power of the sacred impacts intensely on the structure of the human psyche, the person is caught up so strongly in the territory of another world that it causes deep changes in how the psyche is structured. The more secular portions of the psyche are changed and the person experiences this as a "stripping away" of parts of the personality that are irrelevant to a sacred life. People, when subjected to the peculiar stresses and forces of encountering the sacred, when feeling deeply the joys, sorrows, and tragedies of human life, are forced into realms where language falters and fails. At such times, as Orpingalik told Rasmussen: "Songs are thoughts, sung with breath when people are moved by great forces and ordinary speech no longer suffices."[12]

Because ordinary speech is not sufficient, the resolution of conflict or stress through song represents a unique solution to a problem. At the point of resolution, a person is tapping into the particular sonic encyclopedia that exists and is pulling out of it a nonlinear manifestation of a conflict's resolution. When that solution is again expressed in song, it can touch another person who is also in stress, and who can somehow use the song to effect cure.

Songs, like all things, vary in their intensity and power; there are many types of songs for many different purposes. Some songs, as I mentioned earlier, might come from animal spirits who desired to help human beings.

Eagle Shield, a Teton Sioux who lived in the early part of the twentieth century, received many sacred songs for use in healing. All of them were gifts to him from the bear. The bear was considered to be one of the most powerful of all the animals in conveying knowledge to a healer. Those whose knowledge of healing came from the bear were considered to be the most powerful of all healers.[13] As the healer Siyaka said:

> The bear is quick-tempered and fierce in many ways, and yet he pays attention to herbs which no other animal notices at all. The bear digs these for his own use. The bear is the only animal which eats roots

Eagle Shield

from the earth and is also especially fond of acorns, June berries, and cherries. These three are frequently compounded with other herbs in making medicine, and if a person is fond of cherries we say he is like a bear. We consider the bear as chief of all animals in regard to herb medicine, and therefore it is understood that if a man dreams of a bear he will be expert in the use of herbs for curing illness. The bear is regarded as an animal well acquainted with herbs because no other animal has such good claws for digging roots.[14]

One of Eagle Shield's songs from the bear was used with a local Astragalus species. In Lakota, the herb was called *cante' yazan'pi icu'wa*. It was administered in a tea for heart trouble or stomach pain. Before singing it, Eagle Shield would pray:

Te'han mawa'ni kte sni ecan'ni ke'yasi to'ksa eca'na maya'ni ketlo'.[15]

Translation: A long time before I can walk you may think, but (in) only a short time you will be able to walk.

No. 85. "You Will Walk" (Catalogue No. 514)
Sung by EAGLE SHIELD

VOICE ♩=72
DRUM not recorded

Ta - ku wa - kaij ya - tin kte - lo ka - haij - tu - ke

ma - ni ye - lo

Translation: Something sacred you will eat. Now you will walk.

Another herb with which Eagle Shield used song was for diseases of the kidneys. The plant, *Lactuca pulchella* (wild lettuce), was called by the Lakota *azun'tka yazan'pi on'piyapi*. The plant was dried and made into a tea.

Not more than three doses were to be prepared at a time, as it could not stand overnight. Three doses per day would be taken. Eagle Shield noted: "No matter how much a person is suffering, as soon as this medicine reaches the spot it relieves the pain. This is usually done by the time six doses have been taken."[16]

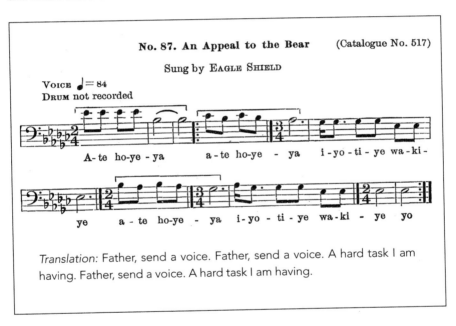

No. 87. An Appeal to the Bear (Catalogue No. 517)

Sung by EAGLE SHIELD

VOICE ♩ = 84
DRUM not recorded

A- te ho-ye - ya   a-te ho-ye - ya   i -yo - ti - ye wa-ki -

ye   a - te ho-ye - ya   i-yo - ti - ye wa-ki - ye   yo

Translation: Father, send a voice. Father, send a voice. A hard task I am having. Father, send a voice. A hard task I am having.

Attention must necessarily be paid to understanding that the song of a plant is a mnemonic device to recreate the sacred state of mind in which the song was first heard. The song by itself, though encoding harmonics particular to the sacred power of an individual plant species, is not sufficient without the understanding and directed intention of the person using it. When a sacred plant song is sung, personal attention is directed to recreating the territory of the sacred that it represents. To mimic a song learned from another without manifesting the underlying experience that generates it means nothing. As the Cherokee medicine man, Swimmer, told James Mooney:

> Merely reciting the formula is not sufficient if we want to obtain success though: we must also know "what is to be used with it" and last but not least we should also know "how we have to work."[17]

One who bought a song had to be sensitive to the spirit world—what was being conveyed was a direct experience of a kind of curing of which the song was its key.

In my own process of learning sacred song, I began with simple talking to the herbs. Eventually I became aware that each plant I had come to know possessed a specific energy of its own. This energy had its own rhythm and eventually began to take on the nature of song. As my bond with certain plants grew stronger, as they began to come to me in waking/dreaming states, I began to notice that I was associating a specific state of mind with each different species. The songs I developed from those times can evoke that state of mind and actualize the power of that species.

In utilizing sacred songs there is a distinct and real meaning to the words Swimmer shared with James Mooney: "We should also know how we have to work." For there are important things to consider, to remember, when one invokes the power of the holy through song, especially when using it to heal.

When preparing to use a sacred song, it is always necessary to pray first. This marks out what is to happen as different in space and time from what has gone before. It says that, from this point on, what comes is a sacred thing, and therefore the internal world of the human is prepared for what follows. When the song is begun it is necessary to attend with all the power of one's mind and spirit. Human beings can discriminate very subtle differences in energy states, and when one is beginning a sacred song, one is crafting a unique and powerful entity that has a distinct purpose. So it is necessary from the beginning and all through the song to attend carefully, checking internally that the song as it is being crafted matches that which has gone before.

Often the plant will lend its guidance in the process, its energy blending with the singer's in subtle harmonies. If the song is being used for healing, it must carefully be directed toward its intended function.

Many indigenous people have remarked in some disgust on the Euro-American's inability to keep the mind focused and directed in such a manner and feel that non-Indians should not work with sacred songs. Though this is too extreme an attitude for me, I too have noted that many contemporary people cannot seem to keep their attention focused to such a directed outcome. Indigenous healers believe that a faltering of will and focus can lead to illness or death. In my own experience, though I do not do so often, at times in which I have faltered I

Swimmer

have noticed a distinct feeling of nausea that sometimes takes quite a while to abate. In all those instances I have worked diligently to repair the damage my failure to hold the power of the song has caused.

When the songs and ceremony are finished, it is once again necessary to pray, to end the sacred space that has been created and thus move once more into secular space and time.

When I express prayer through song, the universe seems to listen in and—in some fashion unknown to me—I and all things around me seem to come together. There is a lessening of the duality that is an inherent aspect of the human condition and, for a time, I and the environment in which I live become one. In this process, when I sing the song of a plant in the rhythms that have come to me, the spirit of the plant comes alive. It rises up out of the slow pattern of life it knows, from its embedding in the life web of the Earth. I, too, shift and together we seem to enter another realm, one in which interspecies communication is possible. In that moment and time there is a facilitation of my ability to speak with the plant about what I need from it and for the plant to hear what I have said. And then, in the process of making medicine, this living spirit that infuses the plant is shaped, formed, focused and goes into the body of the sick person to heal.

I find that the rhythms of the plants (and indeed that of the rocks, the Earth itself) and their individual songs infuse my own life more and more over time. I am rarely lonely, for all of life has its own song and it carries me along in its pattern of sound. I never knew in what a great emptiness I lived until I began singing the sacred songs of the Earth. Perhaps, even more than the surfeit of touch that so many human beings experience, it is the lack of Earth and plant song that causes so much alienation and emptiness.

5
MAKING SACRED
RELATIONSHIP WITH PLANTS

Do you think it possible to dissect a human being,
render it down into its constituent parts, feed them into
a machine which measures such things and determine
from that its ability to paint or create great music?
No? Then why do you think that once you have done
this with my body you know anything about me?

—ANGELICA

*E*ver since those early days on my great-grandparent's farm I had wanted land with which I could develop deep relationship. After many years of work my wife, Trishuwa, and I bought our first land just outside Boulder, Colorado behind those magnificent rock formations called the Flatirons.

When I was younger, I spent three years living in an unimproved seven-by-fourteen-foot cabin in the mountains outside Golden, Colorado. I knew how to live close to the land, to make do with less. But this move had elements that were new to me. This time I owned the land. There was a level of responsibility that had been missing before. And this time, unlike that first time so long ago, we had spent many days in prayer and ceremony asking for land that was strong and alive and holy.

We didn't know what we were getting ourselves into.

Our new house (we were doing all the work ourselves) was but a two-by-six stud-frame box with plywood on it. But more intense than the primitive conditions was the power of the land on which we now were living. It surged up through our feet and into our bones and deep into our spirits. It began turning our world upside down.

Our spirits, so used to the ways of the city, had not developed the strength to contain the power of untamed Earth that was pouring into our bodies. It was a time of upheaval and destructuring as old mannerisms and states of mind were stripped away and we began developing new talents and attitudes more congruent with our new home. But in time we learned how to make relationship with our new home, how to walk in balance on the land. Our spirits and our personalities reformed into a pattern able to hold this surge of the holy through the land and into ourselves. We made relationship with Earth. But somewhere in this process I learned something else: that there is a difference between a wild plant and a domesticated one.

A few months after our move I planted some fruit trees. I thought they would spring up from the ground with no help from me; after all, that is what all the other trees on this land did. But even though the trees I planted were genetically manipulated to be especially hardy stock, and even though I planted them in locations where they were supposed to thrive, they didn't do well. All around them the wild plants and trees grew taller and taller each year. The best thing I can say about the fruit trees I planted is that the deer found them a delicacy.

All the time those trees failed to thrive just seemed wrong to me. How can the life force that enables a giant redwood to grow without human intervention, without fertilizers, pesticides, incessant watering, and wrapping it when it gets cold, fail to work within this other tree that I had planted there? In the ensuing years I received uncounted amounts of unsought advice about those trees whose scrawny skeletons and few paltry leaves still graced the place in which I lived. As the advice mounted it seemed clear that to make them grow I would have to be in service to them, like to a child, and that eventually they would grow bigger and bear fruit. And perhaps then they would grow untended, or perhaps not. But still, it seemed incredible to me that the wild trees could grow so strong and huge without human intervention and these could not.

This led to a deep revelation that has remained with me since that time. There is a strong and powerful life force that moves within the bodies of the wild plants. But I saw that—in spite of the work human beings had done in the genetic development of plant species—what they have created does not carry that life force as strongly. I wonder now, as I did then, how is it that we have come to accept this dilution of life force in what we create as a normal thing?

This was all before I knew of the work of Masanobu Fukuoka, who spent so many years showing how to grow crops without tilling the soil, without fertilizers or pesticides, and still equal the yields produced by technological farming.[1] Fukuoka was born in a small farming village in southern Japan in 1913. He trained in microbiology, specializing in plant pathology. At the age of twenty-five, he experienced a deep insight into the limits of scientific farming and began a lifelong struggle to understand how to channel the force of nature into the plants he grew. Still, there is a great deal of food for thought in experiencing the difference between the wild and the domesticated.

When working on sacred relationship with plants, a person must, of necessity, sit with the plants and come to know them. And as the facility to feel the life of the plants and their communications grows, the difference between a domestic and wild plant becomes even more distinct.

The domesticated plants are beautiful and respond well to human caring but they lack a certain vitality and strength. Many people have remarked on the lack of resistance of domesticated crops to disease.[2] It is possible to feel this weakness in the plants themselves. Perhaps the easiest way to illustrate this is to think of the difference in feeling tone between a wolf and a domesticated dog, or of a mountain lion and the feel of a house cat. This same difference of feeling is also present in plant relations.

There is a great difference between the wild and the tame in making medicine and in sacred plant relationship. The wild plants possess a power and intelligence lost in the domesticated varieties. Further, when one goes to the wild ones, one goes as a supplicant—one *asks*. The person hunting a wild plant is dependent on the wild plant populations and their whims from season to season. With the plants grown on a farm, under the dominion of man, there is a great tendency to think of them as owned property. Because the farmer owns and controls the crop, the balance of power is changed. He no longer goes to the wild and receives the giveaway of food and medicine. The wild now comes to him and grows at his whim and caprice. It is tamed.

Many people have written about relationship with domesticated plants, a relationship where human and plant can create an equal partnership. The Findhorn Garden, the works of Machelle Small Wright, and Masanobu Fukuoka all come to mind. My orientation is different. I want to speak for the wild ones who have never been tamed.

Of primary importance in making sacred relationship with plants is both to experience and treat each of them like a person, as you, yourself, would wish to be treated. All things have awareness and all things are made from the fabric of Spirit. Because of our common birth, from the fabric of Spirit, we can communicate with each other. With nonhuman inhabitants of Earth, the proper form of communication is ceremony.

Ceremony may or may not have words involved but the important thing is the communication of *meaning*. The meaning communicated *must* include a genuine feeling of reverence for the plants and life. Without this, the form itself means little. This is because the plants *know* how you feel; they can feel if you really care or not.

This deep caring is evident in all native peoples and all who live close to the Earth, as illustrated in this example from the Ojibwa:

> The medicine woman . . . brought the writer a plant which she said she had hesitated a long time before showing. Her affection and admiration for the plant itself were evident as she caressed its straight stalk, delicate leaves, and fine white roots, reluctant at the last to part with it.[3]

In beginning your relationship with plants it is important to like them. Perhaps the easiest way to start is to find a plant in the wild that you feel moved by or to which you are drawn. The plants that are most often felt to be emotionally moving are trees and so perhaps this is what might move you.

Go and sit by the plant and just let yourself admire it and enjoy its beauty. Let yourself fall in love with it. And in this process, as you would do with anybody you loved, notice how the plant looks today. Notice how it feels to you. And offer to the plant a prayer or short conversation and ask its permission to sit with it and enjoy its presence. Then notice how you feel. What do you think the response of the plant is?

Many people in western culture do not know how to do this anymore. They are often confused about the difference between projection and the response of the plant. It takes practice to unlearn what you have been taught. It is like the story of the raven who wished to be a white dove and practiced many years imitating a dove, trying to sing, walk and fly like a dove. When at last the raven admitted defeat in this impossible task, he discovered that he no longer knew how to be a raven.

We—all of us—have to greater and lesser degrees been learning how to be something we are not. The structure of our culture has involved a learning process for us as children in which our innate feelings and intuitions, our capacity for sacred experience, have not been actively supported. Instead, we have been taught to rely primarily on rational processes and to view skeptically other means of information gathering.

In this process, in a sense, we have been cut off from our roots to the Earth. A great task facing us is "unlearning" this dependence on rational thought to the exclusion of our other faculties. The unlearning process is necessary to being able to make sacred relationship with plants. A person must learn how to "think like a mountain." Once you begin to think like a mountain, of necessity, you think about the good of all life and human beings take their proper place in that perspective.

Masanobu Fukuoka remarks on this when he notes that:

> The violet growing along a mountain path blooms for no one in particular, but people cannot overlook or forget it. . . . I am fortunate to have grown rice and barley. Only to him who stands where the barley stands and listens well, will it speak and tell, for his sake what man is. . . . [A] university professor told me, "It's best to keep philosophy and religion out of the world of science." If the barley had heard it probably would have answered, "Don't bring science into the world of barley."[4]

It is possible, however, to begin to see the world in this manner. Once the process of unlearning has begun, the old knowledge and ways of seeing the world begin to return of their own accord.

John Seed, the Australian rainforest activist, tells a story about two sisters in Australia who began reclamation of their land by choosing a small corner of a field where native plants still flourished. Starting in the spring, they removed the seedlings of the introduced plants. That is all they did. The native grasses and plants began growing ever stronger and each year the sisters increased their culling efforts. Native plants thought extinct in that area began to appear. When the native plants reached an erosion gully, they took it in stride—the reclamation took on a life of its own. John Seed makes the point that within ourselves this same situation exists. The internal removal of introduced species (beliefs) will result in the native plants (old knowledge) appearing again within us of their own accord.

In this process of unlearning, in the process of feeling and hearing the plants again, one comes to realize many things. And of these things, perhaps stronger than the others, one feels the pain of the Earth. It is not possible to escape it. One of the most powerful experiences I had of this was the year when I traveled to the Florida panhandle.

One day Trishuwa and I decided to go out and make relationship with the plants and offer prayers to them. The place we chose appeared quite lush, with huge trees and thick undergrowth. But as we sat there, a strong anger came from the land and the trees. They had little use for us and told us so in strong language. We spoke with them a long time and did not cower away from their rage and eventually, as we received their pain and anger, they calmed down a little. They told us that we could do our ceremonies if we wished and that they appreciated the thought but that it would do no good. It was too late for that place, it could not be helped, the land would take its revenge for the damage done to it and nothing would stop it. I wondered then how everyone who lived in the area could just go on with their daily lives when this communication from all the local living things was crying out so loudly. I wondered if anyone else felt this rage and anger. A number of years later I read the following excerpt from the book, *Wisdomkeepers*. In the words of Buffalo Jim, a plant medicine keeper of the Mikosukee Seminole:

> The earth is like an animal, Buffalo says. When an animal is sick it wriggles and twitches. Just before it dies it shakes even harder, shakes all over. That's what we call the earthquake and volcano. You'll see, it's already starting to happen. The world is wriggling and twitching and shaking just before it dies.
>
> The Creator made it so Florida was shaped like the nose of a deer. One of these days soon the Creator will break the nose off the deer. Florida will break off and fall into the sea. Yes, you watch, it will happen. The time is just about here. Nothing can stop it.[5]

Unfortunately my experience in Florida was not unique to that state. Wherever I go now, the plants speak of the pain of the Earth and the times to come. As you develop sensitivity to hearing them speak you will begin to hear it too. It is important to understand that as human beings we bear a responsibility for what our species has done. The Earth and the plants understand that. For anyone who wishes to become a carrier of sacred plant medicine, the most immediate task facing them is to receive the pain of the Earth. This includes the pain of all the relations that live on the Earth.

In the beginning this is difficult. The strength of what is happening here threatens to overwhelm. But once you overcome this obstacle, once you can hear the pain of the Earth crying and remain balanced, you are playing a significant part in the healing of the Earth. The first task before a healer, whether working with a person or the Earth ecosystem itself, is to receive the pain that is there. How often it is that the primary need is for someone to receive the pain that we each carry, to be heard. In other times, perhaps, the task for people who carried sacred plant medicine was different. In our time, as spokespeople for the Earth and carriers of sacred plant medicine, we must understand and not hide from the pain of the Earth.

So when you sit with a plant and come to love it and work to hear it speak, it will talk to you of many things. And if it speaks to you of this pain it is important to hear it and receive it. Once the plants know you will carry the burden of your times they will share many things with you, tales of their world and ways in which they may be used for healing—spiritual, emotional and physical. In the process of sharing with you the plants may ask you to do things for them: to offer prayers or tobacco, to carry one of them with you for a long time, to meditate with or sing to them each morning.

If you agree to do these things you must never break your word. The plants do not understand traffic jams, children's school plays, or arguments with your mate. They only understand that human beings have been breaking their word for a long time. To carry the power of plant medicine you must be trustworthy, your word to the plant relations should be inviolate. As in all relationships, trust is built on keeping one's word. The more trustworthy you are the more the plants will tell you and consequently the more power and responsibility you will carry.

Developing Relationship

When first learning about plants it is easiest if you find someone to introduce you to them, and if you focus only on three or four different ones. Each day, spend time with them if you can. Learn all you can about them as medicines and use them for your own. If you think about them often, feeling into them and deepening your emotional relationship with them, they will come to permeate and be an integral part of your life.

Watch them in the wild, observe how they grow in each season of the year. What other plants and animals gain benefit from them? How does their coloration vary from ecosystem to ecosystem? Are their roots different from season to season? From year to year? Do they taste different as medicine from season to season, from year to year? How do they feel in your body? What effects do they have? How does it feel to you to pick a plant and make medicine out of it and be healed by it? How does the plant taste? Do all parts taste the same? How does it feel to your touch? How does it smell?

This attention to detail is a cornerstone of Earth-centered peoples' relationship with plants and it leads to a detailed and minute understanding of plants, their uses, their relationship with human beings, and their place in the ecosystem. This developed body of knowledge, a combination of detailed analytical information and deep emotional/spiritual experience, is the body of power practitioners of sacred plant medicine carry within themselves.

Indigenous peoples have always recognized the two-fold nature of the body of knowledge about sacred plant medicine. They observe and know the plants of their world much better than Western botanists. Their training and observation is minute and they often have complexities of identification more detailed than Western systems.[6] They can often identify and name each plant in their territory in any stage of growth, from seedling to dead leaf.[7]

...ey have a name for every one of the coniferous trees of the region; in these cases differences are not conspicuous. It is clear that the majority of white people are less observant and in many cases know far less about plant life than does the Indian.[8]

In Colombia, the Barasana Indians can identify *all* their tree species without having to refer to the fruit or flowers, something no university-trained botanist can do.[9] And some of the systems of plant identification used by indigenous cultures are more complex than those currently in use by Western botanists. Harold Conklin, in his Ph.D. dissertation (Yale 1954), demonstrated the importance of understanding folk classifications in ethnobotany and showed "there are important distinctions between any two systems of classification. In the case of the Hanunoo, they name many more plants than the Western scientist does, and consequently their botany would be unintelligible without recognition of this basic fact."[10] Such complexity of classification and naming has been noted in numerous other cultures such as the Zapotec that "seeks to relate plants along multiple dimensions!"[11]

In developing this detailed body of knowledge not only your eyes but your feelings are the doorways to relationship with the world and plants. Your body recognizes the plant in ways it can never do with pharmaceuticals and as such you must come to rely on it for information. When you understand how your body responds to plants, then you will know and trust yourself and the knowledge you receive.

It is often a good idea for people desiring to make relationship with plants to carry a plant they feel drawn to in a medicine pouch around their neck, hanging down to heart level, and that they do this for as long as a year. In this manner your body becomes accustomed to the plant's presence and you become accustomed to thinking often of the plant.

At night it is good to sleep with the root of a plant you are working with under your pillow. This often results in easier access to dream medicine about that particular plant and helps deepen personal relationship with it.

When you wish to pick a plant for your use, it is important to offer prayers and tobacco. Prayer and offerings are a primary part of making relationship with plants and honoring their essential nature and character. For all Earth peoples, it is absolutely necessary when conducting sacred plant medicine to pray to the plants and to honor them when harvesting with a gift of some tangible object such as tobacco.

> A small pinch of tobacco is put into the hole in the ground from which it was taken. This is an offering to Nono'mis—the earth, the grandmother of mankind—for the benefits which are derived from her body where they were placed by Ki'tshi Man'ido.[12]

The gift of tobacco is an offering in exchange for that which is taken. It tells plant that the person taking it knows that nothing is free, that there are obligations in all actions. Additionally, the prayers awaken the plant from its sleep, from its embedding in the life web, and let it know that something is expected of it. As the Iroquois David Jack said:

> [T]obacco [is] placed and covered over beside the first plant found. This is done so that the plant will know what you want it for. If a person were to come into the house and grab you and take you away by force you would not know what to do—the same way with a plant.[13]

Prayers are common in all cultures in all parts of the world as is the offering of some tangible thing to the plant. In the Philippines:

> [A] section of the stem furnishing enough bark for only one patient is cut at the same height as the wound and several red or white seed beads are placed at the base of the trunk in payment *(gawad)* to the plant's spirit possessor.[14]

Though it is good to make an offering to the plant and the Earth, it is important that the action be more than rote behavior. There must be underlying reverence and thankfulness when the act is performed. Many people who have heard of offering tobacco do so in a rote manner; unfortunately they carry within them no inherent reverence for the plant, no real understanding that it is alive and equivalent in its essence to the picker. They do not know how to treat it as a human being. But when you truly begin to understand how the Earth and the plants feel and, think you will understand that they do not care for the form so much as they desire sincerity of intent and emotional connection. For us today this is the most important offering we can make to the plant relations and the Earth.

In spite of this truth there is oftentimes disagreement among indigenous cultures, medicine people, and people who have learned from them, over the correct form of offering or prayer. These arguments are generally more indicative of cultural differences than disrespect. As you develop your relationship with the plants, let your forms of offering and prayer develop out of your own relationship with the plants and do not worry about following a rote behavior. Still, in whatever form you develop, look for the deeper reality that is beneath the offering of tobacco and prayers. Because there is a deeper truth here. Irrespective of cultural differences, a prayer and an offering are always made. When and how this is done varies from person to person, from culture to culture.

Another crucial aspect to offering prayer in a manner whereby a deep connection is established is the concept of *Earth medicine*. The power of the Earth flows into each thing to which the Earth gives birth. When a plant is cut, this crucial life force flow is broken and specific intention is necessary to ensure that the power of this Earth medicine remains with the plant and does not leach away. Masanobu Fukuoka touches directly on this.

> The living and holistic biosystem that is nature cannot be dissected or resolved into its parts. Once broken down it dies. Or rather, those who break off a piece of nature lay a hold of something that is dead, and unaware that what they are examining is no longer what they think it to be, claim to understand nature. Man commits a grave error when he collects data and findings piecemeal on a dead

and fragmented nature and claims to "know," "use," or "conquer" nature. Because he starts out with misconceptions about nature and takes the wrong approach to understanding it, regardless of how rational his thinking, everything winds up all wrong.[15]

This knowledge is crucial to maintaining the life force which is Earth medicine in a plant that is being picked for medicine. A great deal of my intention when picking a plant is focused on keeping this life force active in the plant. The only way to actualize this, however, is sensitivity to the existence of this force and the capacity to interact with it knowledgeably. The act of praying and talking with the plant begins this process, but the primary action that *fixes* this power within the plant comes from within the herbalist. The Seminole use their breath but that act itself is symbolic. It is the life force that comes with the breath and is directed through it that activates the medicine. In the case of keeping the life force of the Earth strong within the plant, it is the life force of the herbalist consciously interwoven with the life force of the plant and the Earth that does it.

I call on the Earth medicine and ask it to rise up and at that time the soul of the plant begins to sing. Both the Earth and the plant know of my love for them so they come into me and that essence is held inside my body, and each moment do I attend to it. From then on in the preparation of the plant, in making medicine, in giving it to an ill person, I attend to it. It takes many years to learn how to feel the life force of the Earth in that way, longer still to call on it knowledgeably, longer still to bring it and the plant together, longer still to hold it inside and direct it to some end. For one who wishes to learn, it is necessary to know that such a thing exists and then to diligently work to find the truth.

One of the most powerful ceremonies that I use in sacred plant medicine is that of the sacred pipe. Many cultures used the pipe with plants to enhance their deep connection with plant medicine. Sometimes, before I go to hunt a particular plant, I will hold the image of it in my mind and smoke the pipe, including the plant with me in

the ceremony. At that time, I sc
tobacco and ask its help in fir
additional process helps dee
am hunting, and seems to
But each action in the pr
from time to time and ꜰ
moment and the inform
pick the first plant I fir
on Spirit and the pla
often will do it differently. Som
the use of metal shears when cutting a pl
mind it. As you learn, it is wise to follow the pattern
the beginning. When you are aware of what you are doing and in do
so consciously, you must look for the fabric of the REAL that fits for
you and always check inside and with the plant to make sure that you
are doing honor to you both.

As an example of one approach of offering prayer when harvesting
plants, here is how an Iroquois medicine man, Jesse Cornplanter, picks
plants in a sacred manner:

> In offering tobacco, every detail is explained. So many words go
> with a pinch of tobacco as it is committed to the fire. The message
> goes into the smoke. More words follow and another pinch of to-
> bacco punctuates the invocation. The message of the prayer rises
> to a particular agent. There are all kinds of spirits, but only the
> medicine spirits hear it through the medium of the smoke, because
> they alone are addressed.
>
> If I were to go for medicines I would first burn tobacco and tell
> the plants I was about to gather medicines. Then, all the plants would
> be ready for me to come. The length of the invocation depends on the
> power of the individual priest and the nature of the cause he is plead-
> ing. Immediately, those who are addressed commence listening. The
> tobacco is the medium of exchange which man has and with which

he is able to procure the power of plants and animals; it is the vehicle of communication between men and all the spiritual powers. This is *oye?'gwa owe*—the real tobacco of the *ongwe'owe*, the real people.

Wire grass, for example, must be taken as an emetic only after throwing tobacco on the run and grabbing in passing a handful of the grass which is growing on the path. One must call upon the powers of the medicine to enter one's body in order that one may derive strength from the elasticity of the grass for a particular sport such as lacrosse. One should announce the name of the person who is to use the medicine. Then all the medicine people will attune their ears to the message that they may help a certain person at a certain time for a definite purpose.

Supposing one were hunting boneset to mend a broken bone. The first plant you come to is where you place the tobacco offering on the ground but you leave that plant and go on to a second one of the same species, which you take. This tobacco offering is to the particular plant species and is secondary to burning tobacco to all the medicine before setting out to gather them.[16]

A General Method of Offering and Prayer

As you walk, seeking a plant for medicine, or even before you set out, pray to the plant and ask for its help. Talk to it and tell it you wish to pick it to make medicine. If it is for yourself or someone else, say which. Ask for the plant's help in healing the disease you are making it for. If it is a medicine you are making to have on hand to help others later, tell the plant and ask that it be good medicine for whoever needs it. When you go to pick it, sit with it for a minute first. Touch it and make contact with its body. Make an offering of tobacco; you may sprinkle it on the top of the plants or the ground or make a hole and put it in the ground, whichever feels right to you. When you pick the plant, if it is a root, dig it carefully so as not to harm it. Observe the root. How does it feel to you? Is it glad to be coming with you for medicine? Shake the dirt off.

Bring the root (or roots) home and put it in a place, out of the sun, where you can see it each day. As you pass it during your day, say hello to it, touch and smell it.

Each aspect of this process takes years in which to achieve competence. One must learn how to hold in the mind the spirit and intention of the act in which one is engaged, to direct that intention outward to its specific goal, and to know without doubt when that goal has been achieved. Teachers help in this. Because of their sensitivity they act as a highly refined biofeedback mechanism, allowing students to refine ever further their knowledge of sacred states and processes.

Creating a Sacred Medicine Bundle

I keep medicine bags of herbs I have collected so I can carry them with me for helping myself or others. I have a small one in the car, a much larger one to take to people's homes, and one that is only for sacred ceremony. Sometimes herbs are used for healing without huge ceremony, just a simple thanks will do as you apply it to a hurt that you or another person may have. Other times when the hurt is great and involves spiritual matters, deep ceremony and healing work is necessary.

As you develop your relationship with your plant allies you may want to carry some of them in a sacred bundle for your own use in ceremony and for healing yourself and your family. Each part of the bundle must be as sacred as you can make it, each part constructed with reverence and ceremony. Praying for help from the spirits of the world and the spirits of the plants is often important in constructing such a medicine bundle, as this example from the Karuk Indians of California illustrates:

[H]e prays over the herbs one at a time. He takes up one kind of herb first; close to his face he holds it. Then when he finishes praying over it, then he puts it in the bowl basket. Then afterwards he takes up another kind of herb. He prays a different prayer over it. Then he puts it in turn in the bowl basket. He does that same way to all his herbs.[17]

Sacred
medicine bag

Often when you start such a medicine bundle, objects will come to you as gifts or you will find them in unlikely places. You will live with such a bundle for many years and it will come to hold much power for you and your work. It is important that it be constructed with as much presence of mind as you can generate. Here is an example of a medicine bundle of a Sioux medicine man of the nineteenth century:

> Eagle Shield's medicine-bag was made of the entire skin of an animal called by the Sioux *i'kusana,* "white chin" (mink). . . . In this pouch were placed small buckskin packets of herbs, the large supply being in a bundle which was placed outside the lodge during the day and brought inside every night, being treated with great respect. The number of small packets in a medicine-pouch varied according to the man who used them. Thus the medicine-pouch belonging

Eagle Shield's
mink skin
medicine bag

to Sitting Eagle contained seven small packets, which were said to be only part of the remedies used by him. Eagle Shield had four principal remedies, the most important of which was contained in a beaded case.[18]

In the beginning people feel things about the plants, a kind of intuitive empathy. Eventually, as you develop in your relationship with them, the experience of learning can develop into something more similar to what was shared in chapter three, "Visions of Sacred Plant Medicine." This often takes many years. But as personal strength and knowledge of the plants develops, it becomes necessary to be able to hold in the mind and heart the power of sacred plant medicine without losing emotional balance.

Sioux medicine bag

We must believe without flinching or wavering, we must have a staunch confidence in the power of the formulas. For the spirits we call on "know our minds" and if they find our conviction faltering they will not heed us, nor the words we speak.[19]

In making relationship with plants the most important thing is treating them as equals; the second is having proper ceremony (prayers and offering tobacco) which come from true feelings; the third is receiving the communication of the plants (it must be a dialogue not a diatribe); the fourth is doing what the plant asks of you; the fifth is never violating the bond of trust between you and, if you should happen to violate that bond, making it up (as you would in hurting any close friend) immediately.

6

HEALING IN A
SACRED MANNER

*All treatment of the sick was in accordance with dreams. No one
attempted to treat the sick unless he had received a dream telling
him to do so, and no one ever disregarded the obligations of such
a dream. Each man treated only the diseases for which his dream
had given him the remedies. Thus Shooter said: "In the old days
the Indians had few diseases, and so there was not a demand for
a large variety of medicines. A medicine-man usually treated one
special disease and treated it successfully. He did this in accordance with his dream. A medicine man would not try to dream
of all herbs and treat all diseases, for then he could not expect to
succeed in all nor to fulfill properly the dream of any one herb or
animal. He would depend on too many and fail in all."*[1]

—FRANCES DENSMORE

*A*s my learning with plant medicines progressed I would occasionally return to my family physician for one thing or another.
Increasingly, however, he either did not know what was wrong with me
or if he did he had no effective treatment for it. Further, it was becoming increasingly apparent that, as fine a man as he was, he could not go

below the surface of the disease to its root cause. He did not *know* what was wrong, he simply had information about some of its aspects. If the treatment lay outside of easily administered antibiotics, he had few if any tools with which to treat the disease. I was becoming increasingly dissatisfied with this approach to healing. Finally, after a particularly unsatisfactory treatment, I spent a few weeks in deep thought about the course I was pursuing in the healing of my body. I rummaged through all the old receipts and records of medical treatment I had undergone over the past six or seven years. The picture that emerged was deeply thought provoking. In all that time there was not one disease for which I had sought treatment that I could not have treated more efficiently, more cheaply, and more satisfactorily with the plant medicines I knew. This insight caused me a great deal of perturbation. It resulted, eventually, in a fairly unorthodox decision. I decided that henceforth I would not go to *any* physician for the treatment of disease.

As soon as I made this decision, I began experiencing a multitude of ills. I had a severe pain in my side that lasted for months. My vision began to get cloudy and I had headaches. My stomach hurt. I began to lose feeling in my hands. I had a sharp pain that shot from my right side up under my right shoulder blade.

I became extremely afraid. What foolishness was I involved in? Probably I had cancer, a heart condition, or at the least some degenerative liver disease that would cause me to die horribly. Still, I stuck by my decision.

Over the next year I chased these ills around my body from part to part, using herbs and ceremony for each one. Eventually each would subside only to arise somewhere else. I got to know the organ systems of the body extremely well. I learned hundreds of herbs for treating those systems. I began to face my fear.

I sat in meditation month after month, feeling into my fear. I began to trace it to its origins. And eventually I began to lose my fear. I had been taught, from an early age, that only experts knew what was wrong with my body. When I had finally decided to "take back" that power

I went through a very long period of self-doubt. How was it, I finally wondered, that we have all been taught that the individual human being cannot know what is wrong with his or her body? How is it that strangers have come to be in charge of our healing, the knowledge of our bodies, the knowledge of what is wrong with our bodies, minds, and spirits? I didn't like the answers I was coming up with. I realized that I had been taught to fear death and that these strangers were the ones who had been self appointed to save me from what I had been taught to fear.

I began to enter the world of death and my fear and eventually I began to lose my fear of it. As my fear lessened I began to see more clearly what was wrong with my body, to *know* what was wrong. And in those instances the right plant, the right ceremony became apparent to me. I used them. And I began to feel better than I had since I was a boy. I began to understand the difference between the healer and the disease technician.

When you begin treating others with plant medicines you enter a territory that is as ancient and as unique as that of sacred plant medicine. Not all healers use plant medicines to heal but all healers have certain things they possess in common when they do heal. Perhaps most basic to all is that they focus on healing, not merely reducing the symptoms of disease.

It is common among many types of healers to reduce treatment of a disease to palliation of symptoms. The disease or symptoms are reviewed in current texts, a treatment approach is taken from them and followed in treating the disease. I believe that there is an important element of *need* below this that must be addressed to truly heal. This element is present in many people, felt as an instinctive hunger that sends them from physician to physician for they know not what.

There is a deep-seated need in people who are sick to meet with a healer who can be close and intimate and receive the pain that is being carried. It seems axiomatic among indigenous healers that to heal, one must receive the pain of the sick. When the sick person knows that their pain has been received they can, in turn, receive healing from the healer. I think Theodore Sturgeon touched on this truth when he noted in his short story, "Scars," that:

> There is a time when a thing is a heavy thing to carry and then it must be put down. But such is its nature that it cannot be set off on a rock or shouldered off onto the fork of a tree like a heavy pack. There is only one thing shaped to receive it, and that is another human mind.[2]

Receiving the pain of the sick means that you truly understand the territory of illness the sick person inhabits. Understanding the territory that is inhabited by the sick person, understanding its implications, understanding what it means for the sick person to be there, understanding what is then necessary for the two of you to do together, seeing the pain and crisis of the sick person without flinching, without running from it—this is receiving their pain.

An integral element of healing with plant medicines is entering the territory of crisis, called illness, and being able to remain there without fear. Within one's own body and self, the essence of the healing crisis is held and solved. Then the understanding of what remedies should be used can come, seemingly of their own accord.

> And in fact, in order to extract the evil spirits from the patient, the shaman is often obliged to take them into his own body, in doing so, he struggles and suffers more than the patient himself.[3]

The remedies you then apply come out of your true knowledge of the essence of that disease. They do not come out of a book. *This* is the difference between a technician and a healer.

> Each man must bear his own pain or endure his own fasting if he would acquire power over pain in others.[4]

The illness you find when a sick person comes to you can come from many sources besides disease organisms. It can come from violating one's nature. It can come from injury. It can come from a necessary crisis, important because it allows the destructuring of old ways of thought and being. It can come from negative outside influences. It can come from the anger of the Earth. It can come from failure to pay attention and therefore become infected with another's disease. It can come from war. It can come from loss.

As a healer it is necessary to distinguish the different forms of illness. Treatment of one kind of illness does not work well with another. It is often the case that Western forms of healing do not address the underlying causes of illness. The medicine is applied to the body and the cause that resides in the spirit is untouched. Indigenous peoples made a distinction in all diseases. No two diseases or people were presumed to be identical. They were each treated differently. A Cherokee healer has said:

> The doctor also sang a song each time he treated a person. There was a different song for each disease and the songs of the doctors themselves differed from each other.[5]

Believing that there is a deeper cause to illness than the physical is at odds with the beliefs about healing and disease commonly accepted in our culture today. Like relationship with plants, it is necessary to learn to think in another manner to see deeper than the form of things.

Mircea Eliade noted, "[To] remake a living integrity menaced by sickness, it is first necessary to go back *ad originem*, then repeat the cosmogony." In other words, it is necessary to go back to the original sacred time when all things were whole in order to restructure the sick person. Illness performs a crucial and important function: it allows destructuring of a contemporary state, readying the organism for change. "A state cannot be changed without first being annihilated."[6] It was in the process of going back *ad originem*, and creating original, sacred time that many ceremonies for healing were created.

To do this, to create original time, to be a healer, one should be able to maintain balance in sacred worlds and in the presence of disease. One must be able to give up fear and know without question the limits and extent of personal power. One cannot become a master herbalist until fear of personal death is transcended. One cannot become a master healer until fear of the death of the patient is transcended.

When working with sacred plant medicines in healing the concept of balance becomes of primary importance. Barbara Meyerhoff wrote a wonderful book, Peyote Hunt, about the Huichol Indians and their spiritual traditions. In this book, and a companion article entitled "Shamanic Equilibrium: Balance and Mediation in Known and Unknown Worlds," she explored the need for balance in sacred healing and shamanic practice.

> Shamanic balance is a particular stance. It is not a balance achieved by synthesis; it is not a static condition achieved by resolving opposition. It is not a compromise. Rather it is a state of acute tension, the kind of tension which exists when two unqualified forces encounter each other, meeting headlong, and are not reconciled but held teetering on the verge of chaos, not in reason but in experience. It is a position with which the westerner, schooled in the Aristotelian tradition, is extremely uncomfortable.[7]

She notes that in her early field work, she worked with a Luiseno Indian healer named Domenico. On weekends, when he would see patients, he would climb to the top of his tar-paper-covered shack and stand, one leg tucked up in the crook of the other. It came to Meyerhoff over time that what Domenico was doing was demonstrating his capacity to be balanced in sacred worlds. In her later work with the Huichols, the medicine man, called a *mara'akame,* Ramon Medina Silva, also demonstrated his capacity for balance.

> One afternoon, without explanation, he interrupted our sessions of taping mythology to take a party, Huichol friends and myself,

to an area outside his home. It was a region of steep barrancas cut by a rapid waterfall cascading perhaps a thousand feet over jagged, slippery rocks. At the edge of the fall, Ramon removed his sandals and announced that this was a special place for shamans. He proceeded to leap across the waterfall, from rock to rock, frequently pausing, his body bent forward, his arms outspread, head thrown back, entirely birdlike, poised motionlessly on one foot. He disappeared, reemerged, leaped about, and finally achieved the other side. I was frightened and puzzled by his performance, but none of the Huichols there seemed at all worried. The wife of one of the older Huichol men told me that her husband had started to become a *mara'akame* but had failed because he lacked balance. I assumed that she referred to his social and personal unsteadiness, for he was alcoholic and something of a deviant. I knew I had witnessed a virtuoso display of balance, but it was not until the next day, when

Ramon Medina Silva

discussing the event with Ramon, that I began to understand more clearly what had occurred. "The mara'akame must have superb equilibrium," he said and demonstrated the point by using his fingers to march up his violin bow. "Otherwise, he will not reach his destination and will fall this way or that," and his fingers plunged into an imaginary abyss. "One crosses over; it is very narrow and, without balance, one is eaten by those animals waiting below."[8]

The importance of balance is not limited to the Huichol. In Korea some shamans, usually women, dance balanced on top of sharp rice-straw chopping blades as a demonstration of communication with spirit powers during possession trances. The sharp blades are placed atop a six-foot tower

Korean shaman dancing in spirit possession on rice-straw chopping blades

Photograph courtesy of Edward R. Canda

made by stacking various objects such as barrels, tables, boxes and large jars. This tower represents the vertical axis connecting Heaven and Earth. The shaman, after licking the sharp blades and pressing them into her open mouth, places the blades atop the makeshift tower. There she dances for about an hour, praying to the five cardinal directions and communing with her spirit guides. The proficient shaman does not falter and is not cut. Her ability to maintain literal balance and to transcend usual human limits of physical vulnerability confirm her support by spirit helpers.[9]

The attainment of sacred balance is a necessary and integral part of sacred healing, both with and without the use of sacred plant medicines. It is common to all sacred healers, in all countries and in all traditions. If the healer, in the presence of severe and frightening illness, cannot attain and hold balance, he cannot heal and he and the others present may be subject to infection by disease.

This skill of balance requires an exceptional focus of mind. The emotional impact of the spirit and power of disease, holding to the forefront of the mind the power of the plant medicines, the suffering of the sick person, the requirements of directing the healing ceremony to its successful conclusion, bringing in the power of the Sacred, all these take considerable strength and may have to be maintained over a considerable length of time in the process of healing. One cannot falter or lose balance and the toll it can take is considerable.

The following is an account of the strength of Owl Woman, a Papago medicine woman, and her display of sacred balance in the face of treating disease:

> The phonographic recording of Owl Woman's songs occupied an entire day. She did not wish to sing into the phonograph and insisted that Garcia [her apprentice] record the songs. She sang each song softly in order to recall it to his mind, and toward the latter part of the day she sang with him, but not loud enough for her voice to be recorded. At the beginning of the day, when telling of her visits to the spirit world, she had the appearance of a sibyl,

Owl Woman

with a strange, far seeing look in her eyes. The day was chilly and in addition to the white head covering worn by the Papago women she wore her black shawl wrapped tightly around her, as shown in her portrait. In the first two hours Garcia's interest did not falter and he sang one song after another at her dictation. But there came a time when he left out two or three words. There was much talking in Papago. The old woman was suddenly full of animation and fire. The interpreter said, "She is telling him that he must not be discouraged because he forgot those few words. She says he must go on as if nothing had happened." Garcia rallied to his task and the work continued, but the old woman gave closer attention to her singer. Even to one who did not understand the language it was evident that she was encouraging him and holding his interest. She

Sivariano Garcia

was bright, active, and with an occasional witticism at which they laughed heartily. At the close of the afternoon Garcia was singing steadily with little sign of weariness but her face was drawn and tired, as of one who had been under a long strain. How many nights had she held her singers at their task by the force of her personality, while she watched the flickering life of a sick man![10]

Of necessity, the muscles, powers, and capacities to hold such a personal focus of intention take many years to develop. They are an essential and integral part of holding the force of power needed to heal. In the process, one enters the territory of illness and learns to recognize its landmarks. In the beginning one is often afraid of what is encountered. Illness has a terrifying aspect: death. An important meditation is internally exploring the personal and emotional responses to disease and death that are felt during confrontations with illness. Over time, as one becomes used to it, fear dissipates, the territory becomes more familiar and one comes to walk there with confidence.

For each person this territory appears differently. Its visual and sensory aspects are shaped by cultural and individual idiosyncracies. Each healer interprets that territory as unique sights, sounds, smells, tastes, and feelings. Because all healers are different, each description of the territory of disease tends to vary from healer to healer. But one thing remains certain—they know the territory intimately and without doubt.

This is how Sereptie, a Tavgi Samoyed healer of Siberia, describes his travels in the spirit world:

> When I am looking for a sick man, the road is narrow, like a thread. I do not know who is leading me, in front I see the sun and moon. On (the lower) part of the narrow road there are conical ramshackle tents; on this (road) you go for the breath of the man. The other part of the road (leading upwards) is quite entangled—I do not understand why. The man who is going to recover has a breath like a white thread, while he who dies has one like a black thread. Going along the road, you look sideways and you proceed. Then you find the man's *nil'ti* and take it. (The *nil'ti* is one of the life substances of a human being. It must be brought back by the shaman from the realm of illness.)[11]

For the Huichol it is somewhat different. If a person loses the *kupuri* (the life essence, spirit or soul) then the healer, the *mara'akame,* must go looking for it.[12]

A part of the person, a part of the soul or spirit, often becomes lost and wanders far, trying to find its way back. The causes of this separation are many but without the restoration of the soul, illness continues and gets progressively worse. In many instances the healer must journey to the land of the dead to recover the soul and bring it back.[13] The journey to the land of the dead involves travel to a territory that is described in similar ways in many countries and traditions.[14] There is a low wall or demarcation that is present in the land of illness. Should the soul cross it the person usually dies, for few healers can go beyond this barrier and live. Some of the great shamans of many countries were said to be able to cross this line and return. For others, if the soul resides on this side of the wall, the shaman or healer can retrieve it and bring it back to the person's body. Here is an account of a shaman who crossed the point of demarcation to retrieve a lost soul:

> Accompanied by his helping spirits, he had followed the road that leads to the Kingdom of Shadows. He came to a little house and found a dog that began to bark. An old woman, who guarded the road, came out of the house and asked him if he had come forever or for a short time. The shaman did not answer her; instead, he addressed his spirits: "Do not listen to the old woman's words, walk on without stopping." Soon they came to a stream. There was a boat, and on the other bank the shaman saw tents and men. Still accompanied by his spirits, he entered the boat and crossed the stream. He met the souls of the patient's dead relatives, and entering their tent, found the patient's soul there too. As the relatives refused to give it to him, he had to take it by force. To carry it safely back to earth, he inhaled the patient's soul and stuffed his ears to prevent it from escaping.[15]

All diseases have their particular territory. To heal, the healer must be able to enter that territory, recognize its landmarks without fear, and be able to lead the spirit of the sick person out of that place.

Within the world of sacred plant medicines, the healer acts as an intermediary between the particular sacred power of a plant and the person who is ill. Through ceremony and the capacity of the healer to understand and focus the dynamics of the process, the power of original sacred time is brought into contemporary secular space and time. The spirit of the sick person is taken into sacred space to a time when disease did not exist. Sacred allies are often used in this process to assist the healing. Some of these are plants. The medicines of the plant bodies help the body of the sick person become strengthened and more powerful so that the disease cannot easily remain there. The power of the ceremonies and the plant medicines bring a reformulation to the spirit of the person so that the underlying cause of the disease can no longer exist in that person.

I remember the first day I introduced a person with borderline personality disorder to plants. The client, a woman, perhaps 28 years old, was extremely fragile and unsettled. She was in the midst of a very painful divorce and was experiencing strong rage. During the many times we met and talked, she expressed her feeling that she was empty inside, hollow. She could describe the hollowness, where it was located in her body, how it felt. In this process I had a strong and visceral response about one particular plant that might be of help. So one day we went for a walk and I took her along a stream, deep in the shadows and intermittent sunlight of old forest. Soon we came upon the children of that angelica plant I had met so long ago. I was watching her carefully and saw the impact that plant made on her. She stopped and drew in a deep breath, her body steadied, the constant trembling ceased. The skin across her forehead softened and relaxed and the tension, so long a part of her physiology, departed. Her eyes lost the somewhat rigid fixed staring that had characterized them and became moist and she turned and looked at me and remarked, a small smile playing on her lips, "It's wonderful!"

As we sat, I spoke with angelica in the ways I had learned, introducing her to this woman I had brought to meet her. I shared with my client what I knew about angelica. The woman's hands were in constant motion, fluttering about the plant, touching its leaves as if it were a lover. I asked her to relax and close her eyes and begin speaking to the plant in her mind. The connection for her, as she afterward reported, was very strong. The plant seemed to her a tall, strong, mature woman. I asked her then, when she had talked with the plant awhile, to ask it to come in to that hollow place within her. At the moment that she did that, her body straightened, the lines of her face filled out, the little girl look vanishing. When she opened her eyes she said, "For the first time I don't feel hollow and alone inside."

In the days that followed I had her practice that exercise many times. She practiced walking and talking and doing her daily work with angelica inside her. I gave her a root to keep with her and some of the tincture of the root to take internally. It helped a great deal. It was not a panacea, many other things were needed, but it did fill this one need. I have used angelica many times with women, often successfully. Many women, I have found, have within themselves a hollow place, like angelica's stem. I give angelica to women who suffer from this or who have an imbalance in their womb (as in reproductive cycles) or in their emotions or in their spirit.

For the healer who works with sacred plant medicine, it is important to be able to recognize the distinct spiritual identities of each plant that is used. This understanding is then matched with the spiritual crisis faced by the person who is ill. The herb, or herbs, are brought into the healing process and their spirit evoked and introduced to the ill person. If done properly the herbs will act as allies for the ill person during the healing crisis. The plants are the friends of human beings and they enjoy helping them.

For the Creek Indians of North America, diseases came into being because humans failed to honor the animals when hunting. The animals, in self defense, created the diseases that now afflict humans. The plants, feeling sorry for the people who were suffering so badly, volunteered the remedies that were to counteract the diseases.[16] For the Cherokee, also, it was similar.

Each tree, shrub, and herb, down even to the grasses and mosses, agreed to furnish a remedy for some one of the diseases named and each said: "I shall appear to help man when he calls upon me in his need."[17]

Within each human being there is the capacity to communicate with all life, to understand the archetype that lives within. The plants will share many of their healing properties, physical and spiritual, to one who listens. First one learns about the plants and their sacred territory, how to relate to them and be honorable. Then one learns how to prepare medicines from them, then how to understand disease and healing. If there is doubt about what course to take or what plant to use, you can pray and ask the plants for guidance.

> When the doctor is in doubt what treatment to apply for the relief of
> a patient, the spirit of the plant suggests to him the proper remedy.[18]

Out of necessity, when one learns to heal, one has to face death and no longer fear it. All human beings die. If one lives beyond one's time, if even an aspect of personality lives beyond its time, one becomes a caricature. Because death is so intimately a part of the human journey one must include it in the process of healing. The healer must have unwavering respect for and belief in sick persons' ability to find resolution to the crises confronting them, even if successful resolution means dying. The territory of illness also includes the territory of death. And death has its own sacred dimensions that must be found and mastered. The healer must have a deep knowledge of the sacred and the plant relations, ceremony, the territory of illness, and the proper relationship of death, and be able to evoke each of these things in their proper time. At the same time, one must know one's place and grow beyond hubris.

There is a power greater than the human that makes all this possible. How many years are necessary for this teaching! It is an ancient archetype of the human expression. Like many things now it has been secularized, split into pieces. But, like the territory of sacred plant medicine, it can be learned. Our ancestors have gone this way before.

7
DIGGING FOR MEDICINE
The Wildcrafting of Medicinal Plants

> *It was Natawika's custom to gather her medicinal herbs*
> *as systematically as a farmer gathers his crops, each in its*
> *season. Many were gathered near her home. The roots*
> *were washed thoroughly and "trimmed up." The larger*
> *roots were cut in half lengthwise, strung on a cord, and*
> *hung up to dry. The small roots were spread to dry on a*
> *table, which extended along the west side of her dwell-*
> *ing exposed to the sunshine. Here she also prepared her*
> *herbs. Plants were hung to dry with the roots uppermost.*[1]
>
> —FRANCES DENSMORE

*W*hen one approaches the Earth and the plants to gather medicines, it is important that it be done with caring and knowledge. Among indigenous groups, people who gather medicine in a sacred manner have done so in similar ways all over the world. Though some of the techniques may differ the underlying attitudes of mind are the same. It is only when plants become viewed as commodities that they begin to be harvested without thought. It can take months or even years to develop the understanding of what you are seeing when you go into the world

Natawika

of plants. There is a world of deep interrelationships that, because we have been so long separated from it, we do not easily recognize. It is important to recognize that the learning takes time and that you come as a student.

Perhaps the most important attitude to learn is that of "thinking like a mountain." This term comes from Aldo Leopold's *A Sand County Almanac*. In his book he shares the time when he first learned to "think like a mountain," when he first recognized the interrelationship of the world.

At that time, he and a friend were eating lunch and saw what they thought was a deer swimming across the stream below them. When the deer finally reached the bank and climbed out they realized it was a wolf. The mother wolf was soon joined by her litter in joyful abandon.

Leopold and his friend emptied round after round from their rifles into them. They both thought that fewer wolves meant more deer to be hunted by man.

When they were finished they had mortally wounded the mother and one cub, the others fled. Leopold noted: "We reached the old wolf in time to watch a fierce green fire dying in her eyes. I realized then, and have known ever since, that there was something new to me in those eyes—something known only to her and the mountain."[2]

In the years that passed, Leopold saw state after state kill off its wolf population and mountain after mountain fall under the voracious and unchecked appetite of the deer herds. There is a relationship between the living things in the ecosystem that, though unseen, holds a great power. When approaching the time of "digging for medicine," you approach that power. And when you approach it, once you learn to see it for what it is, it is automatic to come in reverence, with love and respect. This attitude and understanding is the basic component of seeking medicine and harvesting plants.

Most common diseases can be treated with the knowledge of only ten plants. Some medicine people might only know one or two[3] or they might know as many as a thousand[4] but ten are often sufficient. It is not important to know the plants that come from far away; the ones that grow near to you, in your back yard or in parks near your home, have all the healing power necessary. They, in fact, are better. You all live in the same area, are a part of the same ecosystem. You partake of the same water and air. You are of the same community. When you start with these familiar plants you begin to bond with the land on which you live. You will see and pick the same plants year after year, from the same stands and same land. Each year your knowledge of the healing power of these plants will grow. You will learn many things about them. And as such your capacity to evoke their healing power will also grow.

In each community of plants there are grandfather and grandmother plants from whom that community of plants has come. It is important to leave these plants untouched. Often they live at the top of slopes and seed down. They may be of great age, thousands of years old. To the unschooled eye they appear much like all the others. Some of these plants were here when the great ice sheets retreated north. They saw human beings take their first steps on this continent. One chaparral plant in the southwest desert has been carbon dated to be 12,000 years old.[5] Others are much older. In the presence of such age and wisdom you should come humbly.

The grandfather and grandmother plants should be left undisturbed. They should be honored with tobacco and smudge, prayer and ceremony. When you meet such a one you meet the archetype of its kind and it possesses great power.

Down from the old ones are the younger plants and their offspring. These are to be taken for medicine and food. Offer prayers and tobacco. If you are picking the root, the plant will most often die as a result. It is giving its life for your need—respect and remember this. Some plants, like osha, have very long roots and it is difficult to get the whole root. This is a good thing. When the root breaks off, the remainder, still in the ground, will continue to grow and produce a new above-ground plant the next year or perhaps the year after that. Other plants are not this way and the digging of the root kills that particular plant. For this reason many people feel that roots should be harvested only in the fall after seeding and when the energy of the plant begins to go back below the ground. Some roots, however, are more potent when dug in spring and this is actually the time to dig them. Though most people think plants are dormant in winter, the roots are actually growing the whole time, slowly storing up for spring when they use all that stored energy to produce the new plant and seeds. In spring many roots will be fatter and more vital. Such is the case with osha.

In time as you harvest plants you will come to know which plants are better harvested in spring and in fall. Some roots, such as that of red root (Ceanothus Americanus) are best in the fall after the first good frost. The inner bark of the root attains a brilliant red color and often a pinkish tinge spreads throughout the white root itself. The tincture or tea of the root is deep red in color and is significantly stronger when made from fall roots. Most roots should be dried out of the sun in a well-aired place. It is important to look at them each day, touch them and keep a close contact between yourself and the root.

When you dig the root, replace the dirt in the hole you have made. If the plant is in seed, scatter the seeds. When you dig roots, shake off the dirt against your hand or leg. See and observe the root. Is it healthy? There should be a strong feel of vitality about the root and an absence of rot or mold. Few roots need washing. Coral root *(Corallorhiza maculata)* is one exception. The white coral-like root is so intertwined with rich earth that separating the easily fragmented root from the earth is nearly impossible without washing in a colander. I put all roots in a drying tray. Some people string smaller roots and hang them to dry. Larger roots need to be split in two to prevent mold and to facilitate drying. Roots that are very mucilaginous and wet take much longer to dry than other, dryer roots. If the root you are collecting is a two-year plant, harvest the root in the fall of the first year or the spring of the second year. Roots will usually last many years before losing potency, though for each plant it differs.

If you are picking the above-ground parts of a plant—the leaves and stems—touch them, and enjoy their beauty before you cut them. The best time to pick them is in the early morning when they are still fresh with dew and before the sun has wilted them. If you are going to dry them tie the leaves together in a bundle not much bigger than one inch across to prevent mold. Hang them top down in a dry, well-ventilated place, out of the sun. Make sure that you leave the roots for new plants to come back next year. Know also that in some plants, like goldenseal and American ginseng, the above-ground parts of the plant are just as effective as the roots. The leaf material is not harvested and sold because it does not hold up as well in warehouses. Leafy plants should not be rinsed in water, just shaken and briefly hand cleaned before drying. If there are bugs on them, shake them off vigorously. Above-ground plants will last only about a year before losing potency. Generally, they need to be replaced each year.

Barks are best harvested in the fall though at some periods of the year they do not come off the wood as easily as at others. I tend to cut limbs and leave the whole trees to continue to grow. The bark should be stripped off the limb immediately before it has a chance to dry. It should be dried out of the sun in a well-ventilated place. A few barks though, like white willow, benefit from a few hours of drying in direct sunlight. The smell of willow, so clearly a part of the plant and its medicine, becomes much stronger if allowed to feel the sunlight for a time.

Pay attention to where you are picking plants. Make sure you do not take too many of the plants. Some plants, such as coral root, are quite uncommon in quantity and only one of each four visible plants should be taken. Pay attention to your feelings and take only what you need for medicine. Ask the plants how many you can take. Always pay attention. Pay attention, too, to where you gather plants. In these days and times the land is often ill and consequently so are the plants that grow there.

Whole plants lose their potency much more slowly than broken or powdered plants. Leave plants as whole as you can. Store them out of the sun and air as much as possible. It was traditional in many societies to hang the plants in the rafters of lodges and homes, and use them all winter and then replace them the next year.

The skill of digging medicine is a learned and systematic skill. There are many things to remember and much skill to develop. Many people do it in different manners. It is a craft like woodworking or leatherworking. With the heart you remember the feeling, with the soul you remember the spirit, with the mind you remember the detail of the digging.

Good rule-of-thumb behaviors for wildcrafting are found in Appendix 2.

[W]hen she was a child her mother tried to teach her the uses of herbs, and as she was inattentive her mother stuck a needle through the lobe of one of her ears. She continued careless and her mother pierced the other ear, saying, "This will make you listen to what I say." Her mother told her to try to remember the herbs and their

Harriet Longley

uses, saying, 'Some time, after I am dead, a sick person who is not being helped by a doctor may come to you and ask for Natawi-ka's remedies." Knowing the importance her mother attached for receiving pay for services, Harriet [told her], "I cannot pay you, but when I use these herbs I will put tobacco in a saucer on the table and I will imagine that your spirit comes and smokes it." Her mother said, "That is the same as if you paid me now." As a result of this instruction and her own observation she was able, in 1929, to gather 56 specimens of plants and trees and describe their uses. A much larger number could have been obtained if the collecting had extended over a wider territory and occupied a longer time.[6]

8
MAKING PLANTS
INTO MEDICINE
The Technology of Herbalism

Taking a pail containing about a quart of hot water, he placed a little spoonful of the three mixed ingredients on the surface of the water at the eastern side of the pail, saying Wa'bunong ("at the east"), then at the southern side, saying Ca'wunong ("at the south"), then at the western and northern sides, saying Ningbi'anong ("at the west"), and Kiwe'dinong ("at the north"). . . . The surface of the water was thus dotted with four small patches of powder. He then opened the tied packet of the principal ingredient, took out one scanty spoonful of the contents, and divided it equally among the patches of powder, placing it carefully in the middle of each, beginning with the east as before but not repeating the names of the cardinal points. The ingredients soon dissolved in the hot water. According to Main'gans the medicine was then ready for use, though some preferred to secure a stronger flavor by boiling it. He said that the taste was agreeable and that the medicine

*should be taken four times a day, the dose to be small at
first and then increased, the full dose being taken in a measure which he gave to the writer. This is made of birch bark
and contains about a tablespoonful. The drawings on the
inside are said to represent animals and to indicate that the
measure was to be used for this particular medicine.*[1]

—FRANCIS DENSMORE

*P*lants are generally taken as medicines in four ways (though this is by no means exclusive): in water as infusions or decoctions; as tinctures from extended immersion in an alcohol and water combination; as salves from transferring the power of the herb to an oil base; and in an unchanged state either by chewing or eating the root, or grinding the plant and taking it directly or in capsule form.

When changing the plant into another form for use as medicine, it is important to say prayers as each stage progresses. If you sit with an alcohol tincture you can notice that it feels different from the whole, fresh-picked plant—the power of the plant is bound up in another medium. Sitting with the plant and calling on its spirit with ceremony and prayer brings it into aliveness and turns it into true medicine.

Making Infusions and Decoctions

An infusion is made by immersing an herb in either cold or hot (not boiling) water. The water should be the purest you can find, not tap water. Water from rain, healthy wells or springs, or distilled water is best. Herbs that have strong volatile oils (those that have a strong fragrance like an essential oil or perfume) are best infused in cold water. Other herbs do well in warm water. They should be left for a period of time, from fifteen minutes to overnight, depending on the herb, to allow the water to absorb the essential elements of the herb. Glass or earthenware vessels are best for making infusions and decoctions. Quart or pint canning jars are very good as they will not break

Main'gans

from heat, and the screw cap keeps the nutrients from floating away in the steam.

An Example of a Hot Infusion

This infusion is used for its general nutritive properties, especially for women in menopause.

Mix together one pound each of dried, cut, and sifted nettles, oat-straw, red clover, alfalfa, horsetail, and spearmint. Put one cup of the mixture in a quart container, fill with hot water and screw on the lid. Leave overnight. In the morning strain the mixture to remove all the herbs and drink the mixture throughout the next two days. Do not keep infusions longer than two days as they begin to go bad.

The rule-of-thumb is to take 16 ounces (two cups) per day of an infusion for someone weighing 130 to 160 pounds.

In general, use the following guidelines to make infusions with hot water:

Leaves: one ounce per quart of water, four hours in hot water, tightly covered. Tougher leaves require longer steeping.

Flowers: one ounce per quart of water, two hours in hot water. More fragile flowers require less time.

Seeds: one ounce per pint of water, thirty minutes in hot water. More fragrant seeds such as fennel need less time (fifteen minutes), rose hips longer (three to four hours).

Barks and roots: one ounce per pint of water, eight hours in hot water. Though some barks such as slippery elm need less (one to two hours).

Cold infusions are preferable for herbs, which perform differently in cold and hot water. Yarrow, for instance, can be quite bitter when prepared in hot water but is not bitter when prepared in cold water. The aromatic components of yarrow, and their corresponding antispasmodic properties, are soluble in cold water while the bitter components of the herb are not.

Cold infusions are prepared in the same manner as hot infusions but each herb will need to be immersed a period of time specific to itself. This can only be learned over time though many herbal books on the market can give guidance (see reference section for suggestions).

Decoctions, prepared with boiling water, can be much more potent than infusions. The general method is to take one ounce of herb in three cups of water and boil steadily until the liquid is reduced by half. Use only a stainless steel or glass container, never aluminum. The dosages can range from a tablespoon to a cup depending on the plant used. Decoctions should be kept only a maximum of two days, refrigerated.

Tincturing Herbs

A tincture for internal use is made by immersing a fresh or dried plant in either straight alcohol or an alcohol and water mixture.

Plants, when fresh, naturally contain a certain percentage of water. A fresh plant is placed in 190-proof alcohol (95% alcohol), one part of plant per two parts of alcohol. For example, if you had three ounces (dry measure) of fresh yarrow it would be placed in a jar with six ounces (liquid measure) of 190-proof alcohol. Mason jars are very good for this. The top is screwed on and the tincture is left for two weeks out of the sun. At the end of that time it is decanted and the herb squeezed in a cloth to remove as much liquid as possible. Alcohol pulls out of a plant all the water it contains. The resulting tincture will be a mix of both water and alcohol. With fresh plants I do not cut or chop them into tiny pieces though many herbalists do so. They feel that the more surface area that is exposed to the alcohol, the stronger the tincture. In my own work, the herbs prefer to be left whole. I rarely tincture roots fresh. When I do, I chop them into smaller pieces. Above-ground plant material I always leave as whole as I can when tincturing. Each person must find what works best.

Plants, as they dry, lose their natural moisture content. Tables are available on the moisture content of many medicinal plants. A good one is offered by Michael Moore through his Southwest School of Botanical Medicine. Some plants like myrrh gum contain virtually none and others, like mint, contain a lot of water. When making a tincture of a dried plant you add back in to the mixture the amount of water that was present in the plant when it was fresh.

Generally, dried plants are tinctured at a five-to-one ratio; that is, five parts liquid to one part dried herb. For example, osha root contains 30% water by weight. If you have ten ounces of powdered osha root you would add to it fifty ounces of liquid, 35 ounces of 95% alcohol and 15 ounces water. Again you do not want to use tap water, for pure water is a necessity. Dried herbs are generally powdered as finely as possible, often in a blender. It is best to store them whole until they are needed. Again, the tincture is left for two weeks and then decanted, and the liquid squeezed out of the herbal material.

With fresh plants you can generally get out about as much as you put in. With dried material, especially roots, you get out as much as you can. Amber jars are quite useful for tincture storage as they protect the integrity of the tincture from the chemical breakdown that can occur from sunlight. So protected, the tinctures can last many years. Herbal tinctures can then be combined together (though a certain few do not combine well) for dispensing. Because of their long keeping quality and ease of dispensing many herbalists prefer tinctures. Dosages for tinctures can be found in many herbal texts (see reference section).

A Combination Tincture Formula for Upset Stomach

Ten milliliters each of yarrow, poleo mint (or peppermint), and betony. Place in a one-ounce amber bottle with dropper. Take $1/3$ to $1/2$ dropper as needed. This mixture will usually quiet an upset stomach or nausea in seconds.

Making Oil Infusions for Salves

The first part of making a salve is to transfer the medicinal properties of the plant to an oil base. Then the oil is made thick and moderately hard by the addition of beeswax.

To make an oil infusion of *dried* herbs grind the herbs you wish to use into as fine a powder as possible. Place the ground herbs in a glass baking dish and cover with oil. Olive oil is a good choice. Stir the herbs to make sure they are well saturated with oil then add just enough oil to cover them by $1/2$ to $1/4$ inch. Cook them in the oven on low heat for eight hours (overnight). Some herbalists prefer to cook the herbs as long as ten days at 100 degrees. When ready, strain the oil out of the herbs by pressing in a strong cloth with a good weave.

To make an oil infusion from *fresh* herbs place the herbs in a mason jar and cover them with just enough oil to make sure that no part of the plant is exposed to air. Let sit in the sun for two weeks. Then press the herbs through a cloth. Let the decanted oil sit. After a day the water, naturally present in the herbs, will settle to the bottom. Pour off the oil and discard the water. Some herbalists prefer to start the oil infusion by letting the herb sit for 24 hours in just a bit of alcohol that has been poured over the leaves. They then add the oil and allow to stand for two weeks. The water and alcohol remain behind when the oil is poured off.

TO MAKE A SALVE

*P*ut the oil infusion into a glass or stainless steel cooking pan. Heat gently on top of the stove. Add chopped beeswax to the warmed oil, usually 2 ounces per cup of oil. Many people like the beeswax grated but I just break it up in a number of pieces and put it in. It melts fine that way. When the beeswax is melted, place a few drops on a small plate and let cool. Touch it—if too soft add more wax; if too hard add a bit more oil. Pour into a jar and let harden uncovered. Use the following formula for a good wound salve:

1 quart olive oil	1 ounce cranesbill
1$^1/_2$ ounce comfrey root	$^1/_2$ ounce osha
$^3/_4$ ounce echinacea	$^1/_2$ ounce usnea
1 ounce chaparral	$^1/_4$ teaspoon vitamin E
$^1/_2$ ounce burdock	4 ounces beeswax

1. Add oil to heavy pot. Do not use aluminum or cast iron. Glass or stainless steel is best.
2. Grind all herbs into fine powder, or as close as you can get.
3. Add herbs to oil.
4. Cook overnight in oven with low setting (150 to 200 degrees).
5. Take out and let cool, then press herbs in cloth to extract all oil.
6. Clean pot; return oil to pot and reheat slowly on stovetop.
7. Measure out about 4 ounces of beeswax (generally 2 ounces of wax to every pint of oil).
8. Add vitamin E, about $^1/_4$ teaspoonful, and stir in.
9. Pour into salve containers and label.
10. You may add essential oil if you want to give a fragrance to your salve.

Some wounds do not respond well to a wet dressing like a salve. In that case I use the powdered herbs directly on the wound. The herbs used in the wound-salve formula, when ground into a fine powder, stop bleeding and facilitate rapid healing while preventing infection. After the wound has begun to heal, wound salve continues that process.

The comfrey root facilitates rapid cellular healing and wound closure with less chance of a scar. Echinacea, usnea, chaparral, and osha provide antibacterial, antifungal, and antiviral properties. Burdock is an excellent skin remedy and cranesbill stops bleeding.

Using Whole Herbs

The same herbs used in the wound salve are excellent when placed directly on a wound. They are powdered as finely as possible, combined together, lightly dusted on the wound. The fine powder helps reduce friction against the wound that sometimes occurs from larger, rougher grinding and it can be added to socks and shoes to treat athlete's foot.

Many herbs can be eaten when needed. Osha is a prime example and can be used for sore throats and upper respiratory infections of both viral and bacterial origin. It is very strong and a little is eaten when necessary. Sometimes a combination of both whole herbs and tinctured herbs works well.

Herbal Regimen for an Ulcerated Stomach

[In the treatment of ulcers] chemical drugs would be used that reduce the production of stomach acid and so reduce irritation of the stomach's mucosal lining. Tagamet is a commonly used preparation of the drug cimetidine, which inhibits gastric acid and pepsin production in the stomach through blocking histamine receptors. This makes the drug a widely used treatment for duodenal and gastric ulcers. It is now the most prescribed drug in North America, followed by Valium.[2]

—David Hoffmann

Until recently, ulcers were thought to be caused by an overabundance of stomach acid brought on by stress, diet, or other environmental factors. Physicians attempted to control the excess acid by the prescription of drugs that would lower the amount of stomach acid being produced. A recent discovery by a researcher—that most stomach and duodenal ulceration was being caused by the presence of a hitherto unsuspected bacterium, *Helicobacter pylori*—brought that treatment protocol into question. Acid blockers, before this discovery, were generally prescribed indefinitely or for lengthy periods of time; the initial cost was approximately $150 with an additional monthly cost of $40. Patients were supported in radical diet changes and techniques designed to reduce stress. In spite of this, discontinuing the use of acid blockers quite often resulted in new ulceration. New protocols explored in a study conducted by the National Institutes of Health in Maryland, however, indicated lower long-term costs and significantly reduced treatment periods.

Researchers examined three medical approaches to the treatment of ulceration assuming that *Helicobacter pylori* was the main cause of the problem. Without confirming its presence through other tests, the cost of immediately eradicating the bacterium, using antibiotics and acid blockers, was $372. The second approach, using an endoscope inserted through the gullet into the stomach to confirm the presence of the bacterium, entailed a cost of $391, although the cost of endoscopic tests can run from $598 to $3000 in additional fees. In the third scenario, costs escalated to $679 if the physician followed the old procedure of prescribing acid blockers, and then prescribed antibiotics only if the ulceration recurred.[3]

Immediate treatment with herbs, assuming the presence of the bacterium, is approximately $80, cheaper than all three methods explored by the medical researchers in Maryland. A comprehensive herbal treatment for severe stomach ulceration would involve the following herbs: marshmallow, comfrey root, bismuth, grapefruit seed extract, *chaparro amargosa*, and desert willow. The herbal regimen would use the following protocol:

1. Powdered marshmallow and comfrey root combined in equal parts. Two tablespoons, twice a day, mixed in apple juice, for 30 days duration.

2. Powdered marshmallow root alone, two tablespoons in apple juice, twice a day, for the next 60 days. Note that both the marshmallow and comfrey should be taken in powdered form only, not encapsulated; they are able to completely coat the stomach only if in powdered form.
3. Three hundred mg. bismuth three times a day for the first 30 days.
4. Six drops grapefruit seed extract, three times a day for 15 days.
5. *Chaparro amargosa* and desert willow extracts, combined in equal parts. One teaspoon of the liquid extract, three times a day for 15 days.

The grapefruit seed extract and chaparro/desert willow combination kill the bacterium that causes ulceration. The marshmallow root, comfrey root, and bismuth soothe and heal the ulcerated stomach.

This treatment regimen is quite effective; people have reported noticing improvement in as little as one week, even in cases of ulceration that have regularly recurred over twenty or thirty years. As a word of caution: some medical researchers and herbalists feel that the high concentrations of pyrolyzadine alkaloids in comfrey root present a danger of liver damage. Because of this, I generally use comfrey only if the damage from ulceration is severe and I limit its use to 30 days.

The controversy surrounding comfrey brings up an important point regarding the many traditions of herbal use. There is a lot of controversy in all areas of herbalism as to how to use herbs, which are safe, and who should be able to use them. Many physicians feel that herbal approaches belong to a simpler, unscientific period of human development. Many medical herbalists believe that only highly trained practitioners should recommend the use of herbs, and the herbs used should be laboratory certified for the presence of active constituents. Many traditional Chinese medical practitioners feel that only people trained in that approach should use Chinese herbs. Traditionally, however, herbs have belonged to the common people. They were grown near their homes, found in fields, and prepared simply for use among family and friends, without the intervention of a professional class of healers. This last point is a crucial one in meeting the health needs of a burgeoning world population.

The World Health Organization and UNESCO, attempting to address the problems of overpopulation and health, commissioned a study on this in the mid-1970s. They found that to meet the health needs facing an increasing population, systems of treatment other than Western, technological medicine must be used. In May 1978, as a result of this study, the World Health Assembly adopted a resolution "which requested the Director General of WHO to initiate programs designed to evaluate and utilize 'traditional medicine' to meet world health needs by the end of this century."[4]

Many of the traditional approaches to healing utilized are quite different from those advocated by Western science. Many of them are founded on different assumptions, values, and sources of knowledge. One of these is sacred plant medicine. There is a danger, in trying to codify and scientifically authenticate such widely divergent approaches to healing, that the essence of approaches such as those in sacred plant medicine will be lost.

To make the acquaintance of an herb, to understand the lowly weed, to hear its voice and that of Spirit teaching how to make it medicine and use it for healing, is the essence of Earth relationship and Earth healing—the essence of herbalism. It should be a simple thing unencumbered by experts licensed by the state. It belongs to the realm where the human and sacred meet in the plant. That is where it should belong. As a result all healers will find their own way and will always be different. This is as it should be. We who go before only point the direction. Among the Ojibwa it is said:

> The remedies were individual, not general, and an individual, when questioned, invariably replies, "I can tell you about my own medicines. I do not know about other peoples' medicines nor their use of the same plants."[5]

9
CEREMONY
Making Deeper Relationship with Plants

W hen I was growing up my mother's father insisted that I and my siblings be trundled off to church periodically. I found the experience inexplicable, empty of meaning, and without essential humanity. In short, it was about the most boring thing I had ever been forced to do, surpassing even school. I lived instead for those times when we gathered on my great-grandparents' farm. Somehow in that time, a living ceremony occurred more akin to what I felt church should be about. No one spoke of it but we felt it just the same, and I came to identify worship as possessing a strong current of love and a living essence, a thing missing from the church services I was forced to attend.

Over the years, as I found teachers, as my visions unfolded, I began to work with real ceremony, ceremony that possessed that current of love and living essence. For I found that the Earth cannot understand language, that most powerful of human inventions. Embedded in the body of ceremony is the language that Earth understands.

Ceremony may be self derived, it may come from vision, it may be given by a teacher, it may be cultural. But from all sources it has the same underlying root. It is a process in which the human capacity for sacred feeling and reverence is given form and expression. One tells the Earth, one tells Creator, what is felt and thought through specific actions and movements and intentions. This underlying intention held within the ceremonialist is given outward expression in ceremonial form. And in the process humans, the spirit world, the different elements of Earth are bound together in a living fabric that is alive, vital, and new.

It has long been known among indigenous cultures that when people forget their place in the web of life without periodically renewing connection with the sacred, illness and disharmony follows.

John Seed notes that the Hopi, living in the longest inhabited settlements in North America, still don ceremonial masks and perform community ritual designed to restore connection to the Earth. He makes the point that this connection must be frequently reexperienced during ritual if it is not to be lost. This propensity to lose connection is probably a normal and natural part of our human makeup. Connection must be renewed through ceremony over and over again.

Many ceremonies are simple, for often it is not the form of the ceremony that is important but its intention and meaning. It is a simple thing to thank Creator for life each morning, to ask for the help of the plant relations, to smudge with sage. Other times, it is a good thing when one sees a place that holds sacred power to go to it and say thanks, to leave an offering, to honor the Earth.

Because the interactive process of honoring the Earth and all life has largely been forgotten and abandoned by many people, each one who begins anew to act in this sacred manner helps restore the life web and the proper balance of the Earth. Sun Bear, the Ojibwa teacher, often mentioned that in the coming years of Earth changes, only those who were in community would survive. The survivors would not be those who were sitting over their can of pork and beans with a shotgun. At first I thought that he meant human community; but as I thought more on this thing, I have come to understand that it means each of us must learn to sit again in the community of the life web as an equal, in balance.

There have been many primary ceremonies and rituals developed to help people once again be in balance with the web of life. Among the North American indigenous cultures some of these are the sweat lodge, the sacred pipe, vision quest, and the medicine wheel. In working with sacred plant medicine, I have found two of these to be of especial power and benefit—the sacred pipe and the medicine wheel. I have chosen to share about these two for specific reasons.

The sacred pipe ceremony shows how detailed and sophisticated ceremony can be when filled with the directed intention of spirit. Further the pipe, like many Asian ceremonial processes, is specifically focused on the act of uniting Heaven, Earth, and Man. This unique joining has particular relevance to sacred plant medicine and a great deal can be learned from it. Finally, the use of the pipe was moderately common among practitioners of sacred plant medicine. One example of this use is included in the chapter detailing the use of pasque flower.

The medicine wheel incorporates an essential element of sacred plant medicine, the indigenous understanding that all of life is a circle, that each element of the life web sits together in a council of life, and that human beings travel around a great wheel as they progress through life. Thus, the medicine wheel has particular relevance because it is a highly developed expression of the Earth-centered experience. It incorporates the belief that there are unique and specific stages of human develop-ment that transcend psychology and environment, and that plants have a specific place in helping at certain stages of travel around the wheel. Finally, herbs sometimes contain within themselves the power of a spe-cific direction, and awareness of this power when working with plants can be of special benefit during healing ceremonies.

Each of these ceremonies can utilize smudging: the sacred pipe always, the medicine wheel sometimes. So before I go into detail on these two ceremonies, I will talk a bit about a central component of Earth ceremony: smudging.

Smudging

Smudging is the act of burning an herb and "washing" one's self, the other participants, and the plants or ceremonial tools in its smoke. The act of smudging demarcates the ceremonial event in time, saying that from this point on what we do is sacred. Historically, one of four herbs have been used for smudging: cedar, sweetgrass, sage, or wormwood.

When smudging, the dried herb is placed in a bowl or other container and ignited. The flame is then extinguished, allowing the herb to smolder. The smoke that rises is then fanned on the object or person being smudged, using one's hands or a feather.

The herbs traditionally used for smudging are thought, in all cultures in which they are used, to clear negative influences and restore balance.

The Sacred Pipe

The American Indian ceremonial pipe has been in use for centuries. The earliest pipes that have been found are simple tubes discovered in prehistoric mounds in Ohio. They are made of clay and alabaster. The earliest long stemmed pipes had bowls made in the shape of animal heads.

The use of ceremonial pipes spread along the Mississippi River and into the lakes and eastern plains around the 11th century. One of the first tribes observed by Europeans to use the pipe were the MicMac of Nova Scotia. The pipes have been used in holy ceremonies by Earth-peoples for at least a thousand years.[1]

—JOHN FREESOUL

During the many years I have worked with the pipe, I have followed teachings given in visions, studied with others, read what was available, and learned what I could of its history and tradition.

Most indigenous peoples of North America who use the pipe have legends of its creation or appearance in their culture. Some tribes received it as a gift from another tribe, others received it from holy people or supernatural beings who were sent by Creator to bring it to humans. This gift of the pipe by the intervention of Creator through a sacred being or prophet is akin to the tablets given to Moses in Hebrew tradition, or the appearance of Christ in Christian tradition.

An example of such a story of the appearance of the pipe is this ancient one contained in George Catlin's notes from his travels among the tribes of North America:

> At an ancient time the Great Spirit, in the form of a large bird, stood upon the wall of rock and called all the tribes around him, and breaking out a piece of the red stone formed it into a pipe and smoked it, the smoke rolling over the whole multitude—then he told his red children that this red stone was their flesh, that they were made from it—and that they must all smoke to him through it—that they must use it for nothing but pipes: and as it belonged alike to all the tribes, the ground was sacred, and no weapons must be used or brought upon it.[2]

Each part of the pipe represents certain things. The bowl is the female, the stem the male. The flesh and blood the bowl, the bones the stem. The bowl the Earth mother, the stem Father Sky. The channel through the pipe stem and bowl represents the direct connection between all things and Spirit, the straight and narrow path each human walks to be in relation with Spirit. Thus, when the pipe is joined, all these things are joined as one. Duality is gone.

The tobacco that is smoked in the pipe is made from a variety of plants. There is often tobacco *(Nicotiana)*, mullein *(Verbascum)*, uva-ursi *(Arctostaphylos)*, sage *(Artemisia)*, raspberry *(Rubus)*, red willow *(Salix)*, and sometimes many other herbs. Whenever possible, the pipe carrier must pick the herbs. The plants should be picked in a specific sacred manner and prepared as smoke mixture by the pipe carrier. Prayers of thanks are offered to the plants who give themselves up to be smoked.

When smoking the pipe, pinches of tobacco mixture are offered to the four directions, the plant kingdoms, the animal kingdoms, the elementals, Great Spirit, Creator, the Earth, the sky, humans, sun, moon, all those in Spirit, the rocks of the Earth, and sometimes many more. In this way, when the pipe is smoked, the whole universe participates.

With each puff of smoke a prayer is offered to be carried up to Creator, Great Spirit. The pipe is used in prayer for whatever is needed. There are some guidelines in addressing the integrity of these prayers—that the prayers honor all life, human and nonhuman, and the continuing existence of all relations, not just human. In this way the pipe is a major Earth-centered tool because it belongs to all life, not just humans. It represents and speaks with the life essence in all things. In this way, the pipe carrier is not the owner of the pipe but its keeper. The pipe carrier is a spokesperson for all life and the life force of creation.

In the story of White Buffalo Calf Woman told by Black Elk to Joseph Epes Brown, White Buffalo Calf Woman spoke of this when she said:

> With this sacred pipe you will walk upon the Earth; for the Earth is your Grandmother and Mother, and She is sacred. Every step that is taken upon Her should be as a prayer. The bowl of this pipe is of red stone; it is the Earth. Carved in the stone and facing the center is this buffalo calf which represents all the four-leggeds which live upon your Mother. The stem of the pipe is of wood, and this represents all that grows upon the Earth. And these twelve feathers which hang here where the stem fits into the bowl are from Wanbli Galeshka, the Spotted Eagle, and they represent the eagle and all the wingeds of the air. All these peoples, and all the things of the universe, are joined to you who smoke the pipe—all send their voices to Wakan-Tanka, the Great Spirit. When you pray with this pipe, you pray for and with everything.[3]*

---

*Reprinted by permission of University of Oklahoma Press

There are specific pipe ceremonies, depending on the lineage and culture of the pipe carrier, and the visions that have given rise to specific pipe ceremonies number in the hundreds or, perhaps, thousands.

The pipe can be used for communion with Spirit, help during difficult times, or for healing the sick. How it can be used in these instances varies tremendously depending on the need or use of the moment. Pipes are made from many materials besides the well-known pipestone; any soft stone will do. Alabaster and soapstone, for instance, are often used. The stems can be made of any wood that feels right. If one is called to the pipe, the correct form will soon present itself.

As the pipe carrier engages in the pipe ceremony over time, a deep relationship with the pipe is established. Eventually the pipe "awakens" and comes alive. An awakened pipe is a channel between the pipe carrier (the human community) and Spirit. Just as there is a spirit and soul in all things so, too, is there one in the pipe. The pipe carrier establishes relationship with the spirit of the pipe and through it acts as a sacred intermediary for the human community, the Earth, and for all life. Frances Densmore makes a reference to this event in her article "The Belief of the Indian in a Connection Between Song and the Supernatural." She notes:

> On one occasion the writer was questioning Lone Man, a trusted Sioux informant and singer, concerning information received from a pipe. He was asked whether a spirit had entered into the pipe and gave the information. He replied that this was not the case, saying that under certain conditions a pipe might "become sacred" and speak to the Indian.[4]

This living spirit of the pipe is and should be consciously evoked and related with by the pipe carrier when working with the pipe in ceremony.

Almost all Earth-centered practitioners have some physical object of the Earth that is used for ceremonial purposes. These objects become a focus for communion with Spirit and an honoring of the soul and spirit essence in all creation. They often play a special part in Earth relationship and rites of passage for humans. It is a tool for balancing between the worlds and developing depth of Spirit for those on the Earth-centered path.

Each part of the pipe ceremony represents communion with a sacred archetype of the universe. As years are spent in this communion one comes to understand these archetypes and to be in personal relationship with them. Through them one approaches closer to Creator and one's own true nature. The special power of the sacred, which comes from this process, can be evoked to help one's self, friends, family, and community during troubled times.

When people work with sacred plant medicine they are consciously evoking the power of Heaven and Earth by calling on Creator in their prayers. Within the body of the practitioner of sacred plant medicine, the two are united in balance and sacred power is evoked when the plants are called on to become medicine. When one uses the pipe this process is enhanced. The pipe is akin to many martial art forms that have their origins in Asia. These martial arts have at their center the act of joining Heaven and Earth in the body of the practitioner. This act is a unique one and central to any deep knowledge of sacred plant medicine. In the following detailed account of a generic pipe ceremony, the specific mindset necessary to each ceremonial stage is discussed. Each act deepens the joining of Heaven and Earth and then goes on from there to direct intention toward some expected outcome. This process is central to all forms of ceremony.

To truly understand ceremony and enter into that world, one should come to understand that certain states of mind, of power, exist for which there are few words in the English language. But indigenous tribes, closely connected to the sacred as a part of their cultural life, had well-developed terms for these things. This is especially true of the pipe.

Terminology of the Sacred

In all tribes, each had a specific term that was used to refer to the sacred. The term among the Sioux was *wakan;* among the Ojibwa, *manido;* among the Omaha, *wakanda;* among the Mandan, *ho'pinis.*[5] These are similar to the Asian *tao.* The act of limiting those terms to mean "sacred" is somewhat incorrect, however. It means much more than that. The word refers to the sacred center of all things from which all things have come. So, using Sioux terminology, and quoting Joseph Epes Brown, the degree of *wakan* a thing possesses "is in proportion to the ability of the object or act to reflect most directly the principle or principles that are in *Wakan-Tanka,* the Great Spirit, who is One."[6] The commonness of this concept among indigenous cultures in the Americas indicates a general sensitivity to perceiving manifestations of the sacred and a refined capacity to distinguish the degree of sacredness a thing possessed. As this characteristic was so fundamental to all things in indigenous life, understanding of the pipe or any other sacred object or ceremony would be impossible without it. Within each origin tale of the appearance of the pipe to native peoples, these particular terminologies are present, the meaning of which is culturally implied; thus, it is imperative to understand that which is not said as well as that which is said.

The Ceremony of the Sacred Pipe

The pipe as a sacred tool combines the most powerful elements of two processes. First it is equivalent to the Buddhist meditation process of working with a *koan.* Koans are statements designed to force one beyond the rational meaning of words into an awareness of a higher truth. One of the most famous is: What is the sound of one hand clapping? The pipe, when worked with devoutly can, like a koan, stimulate one beyond this everyday normal reality into an awareness and understanding of deeper sacred truths. Second, the pipe is an act in which the inherent duality of the universe is made one.

Each act in the pipe ceremony represents a specific sacred meaning. In other words, there is an underlying discrete and specific spiritual meaning that is evoked at that moment in the pipe ceremony. Though there are differing elements in pipe ceremonies, depending on the lineage and type of pipe being smoked, generally all pipe ceremonies are identical in essence.

Besides the pipe (bowl and stem) there are usually a number of other objects that go up to make a pipe bundle. There is the smoking mixture in its own container and a pipe tamper—usually a stick that is narrower at one end, used to pack the pipe before it is lit and during the smoking itself—and matches. There is also smudge, usually sage or wormwood; a container in which to carry it; a smudge bowl; and often a feather or bird wing to be used to waft the smoke onto the objects being smudged. There is usually an altar cloth or small rug on which everything is placed, and a container in which all these things may be kept. Then there are any other sacred objects that may be important to the pipe carrier and all of these, together, are carried within a ceremonial pipe bundle.

When the pipe bundle is unpacked the altar cloth is laid down and the objects are laid out. Smudge is lit and everything is smudged, beginning with the people involved in the ceremony—the pipe carrier first. The pipe is usually smudged last—the stem first and then the bowl. When the stem and bowl are smudged many pipe carriers pass them across the smudge bowl north to south then west to east. The stem first and then the bowl are held, their opening to the lips, and smudge drawn through them so that they are purified inside and out. In this beginning element of the ceremony the mind has a chance to quiet itself, the participants to take on the proper attitudes and states of mind. This signals that what is to follow is a sacred thing, set off from what has gone before.

The next step is the joining of the stem and bowl. The bowl is held in the left hand, the stem in the right and both are held up to Creator and permission to smoke is requested. This is the first act that is koan-like in its nature. The pipe carrier acts as an intermediary for the manifestation of the sacred. It is through the carrier's training, directed intention, and sensitivity that the ceremony evokes all the power of the holy, that it is truly *wakan*. At this point one listens for permission to be granted; if it is, then, and only then, are the bowl and stem joined. At the moment of unity, the duality of the universe is made one. This act is central to the nature and purpose of the pipe. It is through this joining and the participatory smoking of the pipe that divisions, human and nonhuman, between people and cultures, between secular and sacred are transcended. All things become one.

The next step is the filling of the pipe. The bowl rests on the ground, the stem in the lap. Generally four pinches of tobacco are used to fill the bowl. With each pinch some portion of the web of life is invited to come into the circle and into the pipe. All life is brought into the process.

A pinch of tobacco (smoking mixture) is taken between the thumb and first two fingers in the right hand, held up to Creator, and a prayer is said. For example:

> *Creator, Grandmother Earth, all my relations*
> *all the Spirits of this place*
> *I ask that You come and join us now*
> *There is a place for You in this pipe*

A second pinch is taken and held up to Creator, the prayer continuing:

> *All the green relations*
> *spirits of tree, and osha, and all the green growing things*
> *I ask that You join us now*
> *There is a place for You in this pipe*

A third pinch is taken and held up to Creator, and again one prays:

> *All the animal relations*
> *the two-leggeds, the four-leggeds, the wingeds, the*
> *swimmers and crawlers*
> *I ask that You join us now*
> *There is a place for You in this pipe*

And finally, a fourth pinch is taken:

*I call on the powers of the four directions, the
stone people, fire, water, and air
All the star relations, all those who have walked this
path before us
I ask that you join us now
There is a place for you in this pipe*

As each pinch is taken and as each part of the life web is invited to come and join in the pipe ceremony, the pipe carrier is intently focused on the underlying *meaning* that is being evoked. The participation of these other members of our world is actual and literal. One must be able to feel them come and take a seat in the council circle of the pipe. Often many years of work are necessary to truly evoke these powers, to feel them actually come and take their place, to create dialogue and communication.

In the fourth step, prayers are placed in the pipe. If others are participating in the ceremony the pipe is passed to the left. The person takes the pipe and holds it, bowl in left hand, stem in right. Anything that person desires to pray for is voiced. For example: if one needs help with a certain thing, for others or oneself, to offer thanks for the help of Creator, to voice a pain that is carried, thanks for a joy felt, anything, it is offered at this time. The pipe is passed around the circle until it comes back to the carrier.

The fifth step is the lighting of the pipe. The pipe is lit, puffed vigorously until it is going well, then it is held up, stem first, for Creator to smoke. This is another portion of the ceremony that works with underlying *meaning*. The pipe carrier works with personal sensitivity, humility, and intention until the time when one can feel Creator actually come and smoke the pipe. After Creator smokes, the stem is lowered and offered to Grandmother Earth. When Grandmother Earth smokes the pipe it is offered to the Four Directions, one after another, until each one has smoked. Then, and only then, does a person smoke the pipe.

The sixth step is the smoking of the pipe. In this process, a person engages in a deeply interactive process with Creator. The smoke is drawn into the mouth, then into the lungs, and released into the air, whereupon it rises up. The prayers in the pipe take on a visible form. Further, they blend with the body; a portion of this sacred prayer remains in the body and a portion of the body intermingles with the prayer. Then the prayer leaves the body and rises up to Creator, becoming a part of all the universe. In this process the individual becomes joined with all things. The process of smoking is continued until all the tobacco in the pipe is gone.

When the pipe is empty the carrier holds it up and thanks Creator and all things for the pipe:

> *Creator, I thank you for the gift of this pipe*
> *and as I (we) go forward from one day to the next*
> *from one pipe to the next*
> *I (we) say "Yes, yes, yes, yes"*
> *A-ho!*

Then finally, the pipe is taken apart, cleaned, and put away.

Using the Pipe with Plant Medicines

The pipe is an exceedingly complex spiritual tool and it is not possible here to treat it with all the attention it deserves. However, I will share some aspect of how I use the pipe to work with plant medicines.

When I am working deeply with one particular plant and I want to deepen my connection to its spirit, I will conduct a specific pipe ceremony for that purpose, often for many days, weeks, or months in a row. During the pipe ceremony, I will sit in a place where the plant grows and open my pipe bundle before it. Then I will perform the pipe ceremony but focusing always on the plant as a coparticipant. This process facilitates immeasurably the degree of clear communication between me and the plant, and moves me much more directly into the state of mind where the plant can communicate information to me. When that occurs, the spirit essence of the plant will come and sit with me in the circle. And at that time, the plant will take on some physical form, unique to the spirit powers of the plant. Once this has occurred, I will begin praying with the pipe and the plant for help in curing those who come to me in need. As the daily or weekly work with the plant spirit continues, a deep friendship and bond is created. The spirit of the plant is summoned strongly by the pipe but once a bond is created it will come when asked, often appearing for days or weeks at a time. During those times it imparts a great deal of information about its uses, the Earth, and healing. Later, when I work with healing someone, the spirit of the plant will come and, if asked, go into the body of the ill person to help in restoring balance and health. The pipe, for me, is a seminal tool for bringing the plant spirits from a fuzzy and vague feeling into a sharp and well defined spirit presence. This process of plant-spirit definition is, like the other parts of the pipe ceremony, one that takes a while to develop. But even in early stages, it enhances tremendously the power of the medicine of the plant and its spirit during its consumption by someone who is ill.

The Medicine Wheel

Is not the south the source of life, and does not the flowering stick truly come from there? And does not man advance from there toward the setting sun of his life? Then does he not approach the colder north where the white hairs are? And then does he not arrive, if he lives, at the source of light and understanding, which is the east? Then does he not return to where he began, to his second childhood, there to give back his life to all life, and his flesh to the earth whence it came? The more you think about this, the more meaning you will see in it.[7]

—BLACK ELK

There are few permanent shrines in Earth-centered spiritual lineages.

Two of these are the medicine wheel and the sweat lodge. Though the sacred circle is present in all cultures the medicine wheel, a specialized form of the sacred circle, is not. Some version of the sweat lodge is present on all continents and all cultures. Though I will not talk about ceremony of the sweat lodge, I do want to quote from William K. Powers' YUWIPI: *Vision and Experience in Oglala Ritual,* published in 1982 by the University of Nebraska Press. The book is excellent and I highly recommend it to anyone interested in the sweat lodge, vision quest, or yuwipi ceremony. I include it because of his eloquence about the permanence of Earth-centered shrines. The following excerpt conveys perfectly the essence of Earth-centered religious practice:

> Architecturally, the only permanent shrine in Oglala religion is the sweat lodge. It stands, sometimes wavering in the wind, in sharp contrast to the countless Christian churches that dot the reservation—little frame boxes with identical steeples and church bells that look as if they had all been constructed by a mission construction company. All are painted a sacramental white and have blue- or green-shingled roofs. It is as if they had come off an assembly line, just the way federally funded housing projects deliver prefabricated homes intact to the owner's land.
>
> The sweat lodge is a perfect symbol for the Oglala religion; when not in use the structures look rather pitiful: a dome made of willow saplings stuck into the ground, bent over, and tied in place with cloth strips or rope. There is something exceedingly profane about them when not in use, in contrast to the white man's shrines and churches, which are perpetually sacred, set off from the rest of society in a feeble attempt to separate religion from the culture's social, political, and economic institutions. The sweat lodge reflects the Oglala principle of austerity and simplicity: the entire universe is a cathedral: everything is permanently sacred unless desecrated by human foibles that cause disharmony between humans and the rest of nature. At this time a special ritual is required to reinstate

a balance among all living things, and only then are special places like the sweat lodge temporally and spatially separated from the rest of the mundane world; it is only during the ritual itself that special rules of conduct are in force and require different behavior toward nature.

When not in use, the sweat lodge becomes a playground for children, who dodge in and out of the framework, stepping into the central hole where the heated stones are placed during the ritual. It is a stopping place for multitudes of dogs, who lift their legs and declare the sacred saplings, placed there in honor of the various aspects of Wakantanka, their special territory. It is a meeting place for ants, spiders, grasshoppers, and flies seeking refuge from predatory birds who alight on the willow frame during their morning feeding. The sweat lodge is often invaded by a recalcitrant cow or a frightened horse, and it tolerates all these intrusions, along with the constant battering of the wind against its desiccated skeleton. It is partly this tolerance that makes the sweat lodge potentially sacred: like humans, it is subject to the whims of nature and must abide by its relentless impositions.[8]

Like the sweat lodge, the medicine wheel is usually an unimpressive structure. Most are simple circles of stones laid in a pattern on the ground. They are overgrown easily and, if left untended, cannot be found in a few years. But they have an ancient history, from Stonehenge to the 20,000 or so examples on this North American continent. They serve as a pattern for focusing awareness and as a way of relating to all living things.

In the past, when people would encounter a place on the Earth where the sacred was manifesting itself, where hierophany was strong, they would often construct a circle of stones to mark it. In effect, the stone circle said, "This place is sacred. When you enter this place you enter a sacred space and time." In time, more stones would be added to the circle, rising as walls and more clearly defining sacred space. In some cultures, these would be roofed over and become churches. Sometimes, like the ancient shrine of The Mother of the Hunt in Bandolier National Monument in New Mexico, it would become a low wall. In other places there would be only a circle of stones.

The medicine wheel represents the circle of all life. When you sit in the wheel and evoke the sacred, all life comes to sit in council. The human, only one member of the web of life, can use the ceremony of the wheel to restore contact with all the relations of life. The animal relations, plant relations, stone people, spirit relations, all things come to sit in council. Our connections with the world are thus restored and the healing of the Earth begins anew.

The wheel of life, the medicine wheel, is a map for everything in the universe, a blueprint of the web of life. As Black Elk said, this wheel exists in all things. Within each of us, too, the medicine wheel exists as our inner council.

A human cannot sit in relationship with all life if the life inside is in disarray and disharmony. It is important to understand how the medicine wheel exists inside you and to work to establish harmony in all its parts. From this starting place you can then move outward to larger and larger external circles, finally encompassing all of the universe.

The Medicine Wheel inside You

To begin working with the medicine wheel inside you, with your inner council, find a place where you will not be interrupted and that is comfortable to sit. Take a couple of deep breaths and let yourself relax. When you are ready close your eyes and imagine yourself walking along a path in a meadow. It is a peaceful meadow and you can hear the sounds of the birds and the wind in the grasses. As you walk you come to a small rise; you follow it up. As you clear the top you can see ahead of you a small building; the door is open. You approach it and go in. It is one large room. In the center of the room is a large oval table with chairs around it, many chairs. There may be people, animals, or other beings sitting in the chairs and some chairs may be empty. There is a place for you. You sit down and look around the table. These are the members of your inner council, the parts of yourself that are often not attended to in daily life. There is the child, the wise parent, the one who says "no" to everything, the warrior, the sage, and many more. There are rarely fewer than fifteen members of the inner council. Talk to them and make their acquaintance.

Over time it is important that you are in good relationship with them all, that not one of them is outside the circle inside you. This may take a long time. Some of them may be angry with you for not paying attention to them for so long. If you work at it you can come to joint agreement about all actions you take in your life. Then no part of you will be holding out, no part unenrolled in your choices, no sullen part of you hating what you are doing. This results in a powerful and strong wholeness (from the word *holy*) in yourself and in your life and actions. It takes some dedication to achieve it. I usually recommend that people speak with their inner council each morning for at least one year. It usually takes that long to really get to know them all and reestablish good relations.

From this exercise you can see how the medicine wheel can be applied externally as well. Instead of parts of ourselves taking places around the wheel, members of community or family or the life web take places around the wheel. These other beings or people are then worked with in the same way as we work with our internal parts are to establish relationship and balance and harmony.

Working with the inner council is a crucial one in sacred plant medicine. If one travels into sacred domains one is working with the substance of reality, that which makes up who we think we are, what we think others and the world are, and that which really is. This process is a difficult one in that travelers may spend many years examining and restructuring deep elements of their personality. This is necessary because their unconscious parts affect all other acts that take place in their lives. If one fears heights, then one avoids them or feels great fear when height is encountered. Even the thought of a great height can induce fear. To gain power one must of necessity understand the root of this fear and transcend it.

When working with plant medicines, it is often the case that in the beginning it is very hard to distinguish between projected feelings and true information coming from the plant. Intense work over many years with the inner council helps identify what is coming from what source, a skill that is essential in this field. Further, many sacred states are fearful or unsettling. It is necessary to be able to hold one's balance at all times. Strong bonds with one's inner council allow that balance to exist, and further, the members of the inner council act as allies during times of spiritual intensity. In addition, members of one's inner council communicate in their own way with the plants, and richer, more varied learnings are available when one has established the bonds of friendship with one's self.

Creating a Medicine Wheel

The ceremony of building a medicine wheel restores all life by reenacting an ancient rite of making sacred a circle, a holy place to remind all to live in good relationship with all life in this universe. This simple circle of stones becomes a vortex of energy as all life forces are honored in ceremony and join the wheel of life. The restoration and building of holy places in balance and harmony with all life is returning the gift of life given to us by Creator and creation. The wheel is a mandala for the dance of life, the unending spiral of interdependence.

Besides the demarcation of sacred space, the stones also represent the members of the Earth community who come to sit in common council. In the circle of stones there are four gates, each corresponding to a direction of the compass—north, south, east, west. In the center of the wheel is a larger stone that represents Spirit, which is at the center of all things. Each of the directions represents a specific archetype and is often a focus of how that archetype manifests itself in human life. Though there are minor differences between people who use the medicine wheel as a sacred rite, the underlying archetypes and patterns are essentially identical.

The south represents the beginning of life, childhood, the time when spring returns and the new grasses grow. It is the time of hope and the surging vitality of life. The time of renewal and rebirth. At this time one is in contact with caretakers, such as parents or other nurturing beings. One receives.

The west is the time of adolescence when childhood is beginning to pass and one begins to struggle with the demons in one's own soul. This is the time when the warrior within awakens, the time when one goes to the "looks-within-place," that place within all of us where we examine who and why we are. It is a time of going into the dark and finding one's balance and strength; the time when one searches for the unique identity that is one's own; when one strives to identify the meaning and reasons for one's life. At this time people struggle with the stripping away of that which has become unimportant. This destructuring process is often painful and the warrior within each of us awakens to help solve the problems facing us at these times. The end result is personal strength, balance, the capacity to be a warrior if needed, the ability to go into the dark and be unafraid.

From there one moves to the north. The north is the time of middle age, of maintaining the status quo. On the shoulders of those in the north rest the old ones, the young ones sit upon their laps. They are the ones who do the work to help the human world continue. This is the time of caretaking and giving, the time when one can put aside one's self and pay attention to others. The first time of maturity.

From there one moves to the east. The east is the time of enlightenment and old age, when wisdom is the order of the day. The things of youth have been given up and one is concerned about the passing on of wisdom to the young. One begins to have a deeper relationship with the Spirit world and is getting ready to pass over. These patterns occur over and over again in all life, with each thing a human does.

To make a wheel, you must first choose the place for it to be. A quiet corner of your yard is good. Even a quiet place apart in your home would work, though it is better to have it outdoors. When you decide on a place, one that feels good, say prayers to this place, smudge with sage, and declare your intention to create a wheel. Then it is time to find the stones. Go out to places where you feel good, wild places, and saying prayers for help, begin looking for stones: eight larger stones for the gates to each direction, a minimum of four stones between each of the gates, and a large central stone representing the source of all things. Sun Bear's books on the medicine wheel, especially *Dreaming with the Wheel,* offer other approaches to creating a wheel and go into great detail.

Smudge each stone you find, and say prayers, leave a tobacco offering in exchange for taking it. Take it to the place where you are constructing the wheel. When you have all the stones ready, with ceremony, begin the wheel. Ask each stone where it wishes to be placed, then place it there. When the wheel is finished you may add more stones to it later. In the wheel we created on our land outside Boulder, and following the same practice with our new wheel at Vision Mountain in Washington, we gathered stones from all over the world, from as many countries as we could get them, to aid in the construction of a wheel that could help the Earth in these troubled times. Our form of medicine wheel is one large enough so that ceremony can be held inside it, with perhaps thirty or forty people.

Like the sacred pipe, each element of the medicine wheel represents a specific *meaning.* Over time, just like the pipe, it is important to sit in meditation with these different aspects of the wheel and come to understand their deeper significance and be in relationship with them. Once such a level of understanding is developed the particular meanings of the wheel can be evoked for aid in healing and desired directed outcomes.

In working with plants, one becomes aware that there are certain similarities between specific plants and the different sacred powers that are represented in various aspects of the wheel. These plants, in other words, seem to hold within themselves the power of the south or the north. Some seem to hold the whole of the medicine wheel within them. So, when one works with these plants, not only the medicinal power of the plant is evoked but also the power of the specific direction that the plant holds within it. Understanding this can aid when helping a person with a specific plant medicine. Some people struggle with leaving the south, the place of childhood. They resist moving into the west, the place of young adulthood. They resist the development of warrior spirit that is integral to the west; resist encountering the darkness that all humans must face in life; resist learning how to carry their own sword. For someone struggling with this, it can be beneficial to give a plant that is of the west to help the person move around the wheel in that direction. As another example, for persons who have been stressed out for a long time, who have been overly responsible and parental and lost touch with their childlikeness, one may decide to give them an herb of the south to help them remember this aspect of themselves, to help restore balance.

As you come to feel the essence of the directions, you will be able to identify them in the plants you come to know and use for medicine.

10
FOUR SACRED PLANTS OF THE ROCKY MOUNTAIN REGION

Their Sacred Qualities and Uses as Medicine

Pasque Flower
An herb of the south

Osha
An herb of the west

Usnea
An herb of the north

Angelica
An herb of the east

I have learned about perhaps 150 plants and, of these, a few stand out whose spirits are so strong for me that I think of them as spiritual allies. I use them as medicines for physical complaints but also in ceremony for healing the spirit of people who are ill. Here is some of what I have learned about them.

There are four primary plants that have become strong spiritual allies for me: pasque flower, osha, usnea, and angelica. All have distinct personalities and when they come and speak with me they take on a particular physical form. Pasque flower is like a small girl child; usnea a seemingly ageless young man with a wiry head of hair much like the plant itself; osha a male bear, dark brown, grumpy and powerful; angelica a mature and beautiful woman with a quiet and enduring strength.

Each of these plants possesses specific gender for me: two male, two female. And though I do use them with both human sexes, I tend to use them more as male with male, female with female. Brooke's point from the introduction, that plants have "integrated the emotional issues that we two-leggeds struggle with so often," I have found to be true. And in matching gender between plant and human, I have found that a plant can help a person begin resolving emotional issues specific to their maleness or femaleness. Some specifics of this process are covered in chapter six on sacred healing.

The Pasque Flower
Pulsatilla patens

An herb of the south

The botanical name for pasque flower is *Anemone patens* or *Pulsatilla patens*. The Dakotas call it *hokshi-chekpa wahcha* (twin flower). The Lakotas call it *hoksi' cekpa* (child's navel). Blackfeet call it *napi* (old man). Omaha and Ponca call it *te-zhinga-makan* (little buffalo medicine). It was one of the sacred power medicines of the Omahas and Poncas and esteemed very highly. Among the latter two tribes, the right to use the pasque flower was limited to the medicine men of the Te-sinde gens.[1]

The lilac-colored pasque flower is covered with a soft hairy down, usually producing only one or two flowers from its small root. First and foremost it is an herb to calm and soothe the nervous system. It is of benefit when under extreme stress and overloaded with worry, or feeling extremely angry and fearful. This herb soothes and calms and helps release tension. It is food for the nerves. It changes the energetic vibrations of the nervous impulses in the body to ones that are more childlike and happy and free from care but in touch with caring and love. It is specific for use in ovarian pain, and extremely good for neck and head pain, ear and sinus pain.

Though I have not found it to be true many herbalists have reported that the fresh flowers can cause blistering to the skin on prolonged contact. Because of this, the plant was traditionally used as a counterirritant, that is, it is placed on one part of the body, which it irritates, and reduces inflammation or irritation in another part of the body. One such use is the treatment of swollen joints in the hands. Placed on the knuckles it would irritate the skin but relieve swelling in the underlying joints. Large quantities of the tincture can cause vomiting when taken internally, and I have noticed feelings of nausea when taking more than 10–15 drops of the tincture in an hour.[2] I have found this dosage of the tincture to be quite effective in helping alleviate extreme stress or acute pain. Because of the strength of this herb and the possible adverse side effects, it should be used with caution.

In European tradition the anemone flower was said to have sprung from the drops of blood of the Greek god, Adonis, which fell as he lay dying. Adonis was beloved by Aphrodite for his great beauty and she begged that he be allowed to live again. Zeus decreed that Adonis' time was to be divided among himself, Aphrodite on Earth, and Persephone, the queen of the underworld. He was celebrated as a god of plants, his death and resurrection represents the seasonal growth of plants and the regeneration of nature.

The Greek physician, Pliny, noted that the flower was named after the wind *(anemos)* because the flower opened only when the wind blew. And pasque flower does tend to grow only in places where the wind blows.

In the mountains around Boulder, Colorado, the pasque flower is the second flower of spring, coming after spring beauty. By the time the other flowers come up the pasque flower has become like a head of white hair. To many people in many cultures, this plant represents the beginning of life/spring, the youth of humankind, and the full cycle of life in growing old.

This plant has the effect, when taking it as medicine, to calm, soothe and heal the nerves. The sensation is very similar to the feelings of child-likeness and youthfulness that are common in childhood. In the medicine wheel it is an herb of the south, of the newness and springtime of life, an herb of youth.

When an old Dakota first finds one of these flowers in the springtime it reminds him of his childhood, when he wandered over the prairie hills at play, as free from care and sorrow as the flowers and the birds. He sits down near the flower on the lap of Mother Earth, takes out his pipe and fills it with tobacco. Then he reverently holds the pipe toward the Earth, then toward the sky, then toward the north, the east, the south and the west. After this act of silent invocation he smokes. While he smokes he meditates upon the changing scenes of his lifetime, his joys, his sorrows, his hopes, his accomplishments, his disappointments, and the guidance which unseen powers have given him in bringing him thus far on the way, and he is encouraged to believe that he will be guided to the end. After finishing his pipe he rises and plucks the flower and carries it home to show his grandchildren, singing as he goes, The Song of the Twin Flower, which he learned as a child, and which he now in turn teaches his grandchildren.[3]

The Song of the Pasque Flower

I wish to encourage the children
Of other flower nations now appearing
All over the face of the Earth;
So while they awaken from sleeping
And come up from the heart of the Earth
I am standing here old and gray-headed.

Osha

Ligusticum porteri

An herb of the west

The botanical name for osha is *Ligusticum porteri*. Its use by bears is legendary among many cultures and is often referred to as bear medicine. The common Mexican term for the plant, *chuchupate,* is said to be an ancient Aztec term meaning bear medicine. Early pioneers in Colorado called it Colorado cough root because of its effectiveness for that condition. Almost everyone just calls it osha.

Osha is one of the primary herbal medicines you can use if you live in the Rocky Mountain bioregion. Osha grows above 7000 feet throughout the entire Rocky Mountain range, from Mexico to Canada. The plant possesses strong antiviral properties, and should be used at the first minimal signs of flu or cold. It is extremely good for sore throats and bronchial inflammations and will soothe and anesthetize almost immediately.

The tea, powdered root, or tincture is antibacterial and excellent on skin wounds to prevent infection. It is especially good in cough syrups. A simple syrup recommended by Michael Moore in *Medicinal Plants of the Mountain West* is the ground root mixed with twice the amount of honey, steeped for an hour then pressed out when cool. But in any cough syrup recipe, it is excellent.

Osha is also excellent for stomach indigestion. I have found it very useful for cramping and pain associated with the beginnings of ulceration in stomach or duodenum (like its relatives angelica and cow parsnip). But its primary usefulness in disease is associated with its antiviral and antibacterial properties.

The part of the plant used medicinally is its root. The seeds and leaves make an excellent culinary herb similar to chervil. *This plant can be confused with a very poisonous one: poison hemlock.* The primary way to tell the difference is the root itself. Osha's root is quite hairy and possesses a strong smell similar to celery. The inner pith is yellow and has a somewhat soapy feeling. But once you sit with this plant and get to know it, it is impossible to mistake it. Until then, it is important to be certain that what you have is osha. You only get one mistake with hemlock.

The plant's leaves look somewhat like parsley. It can stand two to three feet tall and possesses the characteristic *umbel* (umbrella) as its flower stalk and seed pod. Hence its family name, *Umbellifereae*. In digging the roots a shovel is necessary. They can be stubborn, strong and often grow in aspen groves among the aspen roots, which makes digging it all the more difficult. Osha is present in the mountains from about 7000 feet on up. This is one of the few herbs you can dry in the sun without harming it. It will last for many years dried and will not rot because of the potent antibacterial and antiviral substances in the root.[4]

Osha does not like to be domesticated and is hard if not impossible to grow under cultivation. It is a plant of the wild.

Osha is a plant of the west, the "looks-within-place." This is not an herb that gives itself to others in a kind of caretaking manner. It is an herb for warriors, for those who must go into the darkness and face their own demons. It is primarily a male herb, though like all herbs it is also useful for women. If you sit in meditation with osha, do not make the mistake of focusing on its leaves and flowers/seeds. Its life and power is in its root. The delicate leaves are misleading. So when you sit with it, let your mind travel down along its stem, under the ground to where osha lives. There you will see it, like a bear curled in its den. When you call on osha to become your plant ally you must have your own warrior energy available. You do not beg osha to come and be with you—you ask it from a place of strength and power. When osha knows you are a warrior, too, it will be with you and help you.

For people who are struggling with their own internal demons, who are trying to develop their own warrior strength, for those who fight with the darkness within, them I give osha. You must be willing to become a person of passion and strong feeling to work with osha. You must allow your rage and power to come out, to draw the line in the sand and say, "No more, no farther."

Osha goes to the root of the matter. It is a plant that helps those who are going through destructuring. It understands the stripping away process necessary to deep transformation. It is for those who struggle to learn the "Truce of the Bear."

Much of the energy of osha is bear energy. Bears respond to it as a kind of ursine catnip. They will roll on it and cover themselves with its scent. Males have been seen to dig up roots and offer them to females as part of courting. When a bear comes newly out from hibernation it will eat osha if it can find it to cleanse its digestive system. The bear will chew the root into a watery paste then spit it on its paws and wash its face with the herb. It will then spray the herb over its body. The herb possesses strong action against bodily parasites. How did the bears learn these things? No one knows but the herbal knowledge of the bear is legendary in all cultures. The bear is considered a primary healing animal because it uses plants for its own healing.

Any plant that is considered bear medicine is a primary and potent plant medicine. The bear is the archetype of healing animal and there are many stories about its healing power.

There is one story about a village on Turtle Island a long time ago. An old man came to the village. He was covered with sores and he smelled. He was very ill. He had no possessions and no food.

He came to the first lodge in the village and asked the people there, "Will you help me? I need a place to stay and some food, I am very hungry."

The people drove him off, fearing he would infect their children. They did not want such an old sick man in their lodge. The man then went from lodge to lodge in the village and each time was driven off.

Sick and in despair, the old man came to the last lodge in the village. It was set off a little from the others and sheltered under the branches of a great tree.

The old man approached the lodge and called out, asking for help. A woman saw he was sick and in need. She brought him into the lodge, fed him, and gave him her place to rest. He ate and sank onto the bed and fell into a deep sleep.

The next day the old man was much improved but still too weak to go on. He ate what the woman gave him and rested. Upon waking the next day, however, he was much worse. The woman applied what knowledge of healing she had but to no avail. Each day the old man became sicker and sicker.

At last, having exhausted her healing knowledge, the woman knew the old man would die. He knew it, too, and lifted his hand and called her over.

"I was told in a dream from Spirit," he said, "that there is a certain plant that grows in the forest that can heal me. Spirit told me to tell you of it." The old man described it exactly, saying, "Go and fetch it."

The woman went into the forest and found the plant growing where he had described it, picked it with prayers and ceremony, and returned. The old man told her then of the prayers and ceremonies of preparation and how it was to be made into medicine. She did these things and the old man became well. Each day he grew stronger and stronger.

But before too long he began to get ill again. The woman tried all her healing knowledge and again it was to no avail. The old man just got sicker. Then, again, when he was on the point of death, he called her over and told her of a second dream from Spirit about a plant that could heal him. Again she followed his instructions and again he was healed.

This happened over and over and over again for a year. In the end, after taking the final plant into his body, the old man grew well. He did not get sick again. Eventually he rose from his bed and went to the door.

He turned to the woman and said, "Spirit told me that there was one in this village who was to be taught how to heal the people. I was sent to find you and teach you all I know of healing. I have done so."

And turning once again the old man went through the door of her lodge and into the light. The woman ran to the door and looked out. As the old man passed into the forest he turned into a huge bear and walked on.

In this way healing came to the people. As Two Shields, the Sioux medicine man, noted:

The bear is the only animal which is dreamed of as offering to give herbs for the healing of man. The bear is not afraid of either animals or men and it is considered ill-tempered, and yet it is the only animal which has shown us this kindness therefore the medicines received from the bear are supposed to be especially effective.[5]

Many songs of herbal healing came from the bear. Here is another song from the bear, from the Sioux healer, Oku'te:[6]

No. 57 "A Bear Said This"   (Catalogue No. 518)

Sung by SHOOTER

VOICE ♩ = 80
DRUM not recorded

Translation: A medicine, a root of herb, you will eat. At that place it stands. A bear said this to me.

Osha's taste and smell are strong. Pick its hairy root and feel its power. Taste a bit of it, its taste will explode in your mouth and fill your body. It is a good herb to carry with you in a medicine pouch.

Usnea
Usnea spp.

An herb of the north

In Latin the various species of usnea are called *Usnea barbata, U. longissima, U. hirta, U. florida, U. ceratina,* and *U. dasypoga.* Because of its appearance the common name for usnea is Old Man's Beard. The Dakota called usnea *Chan wiziye.* This has been variously translated as "on the north side of the tree" or "Spirit of the north wind," from *Chan* (on a tree) and *wiziye* (toward the snow and pine trees); the north wind; a legendary white giant of the north.

Usnea is a major herb for treatment of mucous membrane systems such as lungs, intestines, throat, sinuses, and the urinary and reproductive tracts.

Usnea species grow throughout the northern United States. They can be recognized as a grey-green moss growing on fruit trees, fir, oak, and pine. All the species contain both antibiotic and antifungal compounds. They are usually tufted or hairy in appearance. On some trees in wetter areas usnea may hang in long massive strands. In Colorado it generally grows in small tufts at the base of pines and firs. It can be a tuft up to the size of a kiwi fruit and may cover entire stumps and dying trees.

The herb is composed of *two* plants in symbiosis. The outer portion (the cortex) contains the antibiotic compounds and is grey-green in color. The inner portion (the thallus) is visible as a thin white thread inside the cortex when you pull a piece of usnea apart. The thallus contains immune-stimulating substances. The thallus is elastic when wet, somewhat like a rubber band, and stiff when dry. Usnea is round, *not* flat, and is covered by minute projections (papillate). It looks kind of hairy or fuzzy.

Powdered or whole it can be applied to skin infections with excellent results. Tinctured in alcohol, eaten whole, or infused as a tea, it can be taken for internal problems from tuberculosis to acute bacterial infections. As a douche it can be used to treat trichomonas and yeast infections. Herbalists generally use usnea clinically for fungus infections, acute bacterial infection, lupus, trichomonas, mastitis, varicose and tropic ulcers, second- and third-degree burns, plastic surgery, athlete's foot, ringworm, urinary tract infections, colds, flu, bronchitis, pleurisy, pneumonia, tuberculosis, sinus infections, staphylococcus, dysentery, and streptococcus.[7]

Usnea species are very effective in the treatment of tuberculosis. In fact, usnic acid (one active component in usnea) "completely inhibited the growth of [TB] in dilutions of 1:20,000–1:50,000 and weakened their growth at 1:200,000–1:2,000,000."[8] Other sources put the effectiveness of usnic acid at one part per million[9] bringing it into the effective range of streptomycin. From this perspective, effective doses of usnea tincture would range from two to seven droppersful three times a day. Effective treatment would need to last six months.

Usnea has a broad use across many cultures the world over. From general wound healer (Canary Islands), to eupeptic (Italy), to antiseptic (Argentina), to antibacterial agent (Saudi Arabia), to antitumor agent (Chile). Usnea has also been used culturally in delayed menstruation in both Korea and Arabic countries. Its use during pregnancy is contraindicated. Usnea is poorly water soluble so that use as a whole herb or as an alcohol tincture is preferred. However, it has traditionally been used as a tea or infusion so some of the beneficial components are available in a water medium.[10]

Given usnea's prevalence in North America it is surprising that there is little information available on its use by North American tribes. All other world cultures seemed to know of it and use it for a variety of complaints, physical and spiritual. It was used for abscesses by the Kiowa and as a dye by many other tribes.

Usnea represents the north, the place of gray hairs. It maintains the lung system of the planet. When usnea came to me, personified as a young man, and spoke to me of its uses, it told me that its healing qualities are specific for the lung system of the planet—the trees. Its use for people was secondary to this primary function. This was the first time I realized that the plants provided medicinal actions with the ecosystem; that they evolved and developed to help the Earth ecosystem, Gaia, maintain a healthy balance within itself. I realized at that time that it was only because we are a part of the ecosystem that the plants also work for us as healing agents. Usnea helps fight off infection in the trees, and thus serves a crucial function in maintaining rainfall patterns. A potent aspect of usnea's power is that of the power of TREE. There is an ancient compact between usnea and the trees, and coming into contact with the deeper spiritual aspects of usnea, one makes contact with ancient powers that existed long before humans.

As my relationship with usnea deepened and I traveled more into the spiritual territory that usnea inhabits I found that, unlike pasque flower, osha, and angelica, usnea's territory wasn't as easy to understand. It is less familiar to my human sensibilities, the landmarks make less sense, it is more disorienting. I found that to make deep relationship with usnea, I had first to make relationship with the deeper spiritual aspects of TREE and ask permission to have relationship with usnea.

Grandfather Usnea
You stride from the north
With long legs
I call on you
in behalf of all human beings
requesting your teachings

Remind us of the
interconnectedness of all things
Speak to us of the ancient pact
between yourself and the trees
and bring us your healing power

Angelica
Angelica spp.

An herb of the east

The botanical names for angelica species are *Angelica archangelica, A. atropurpurea, A. pinnati, A. grayi, A. tomentosa, A. Hendersonii.* But angelica's common names reveal the sacred uses to which it has been put: high angel, high angelica, archangel, masterwort (wort meaning root or herb from the Old English *wyrt,* i.e., root or herb of the master). The names refer primarily to its angelic qualities. The Creek Indians called it *Notosa.*

Angelica is used by herbalists as a reproductive normalizer; to stimulate delayed menstruation; for cramps (reproductive or intestinal); to normalize digestion and relieve flatulence; as an expectorant during coughs and colds; as a diaphoretic and diuretic to cure urinary tract infections. It is a urinary antiseptic. It has some use in relieving joint inflammations. It will stimulate the body to fight off viral infections, though not so well as its relative, osha. The part used is primarily the root, though the seeds work very well for stomach nausea. The stems and leaves have a weaker action—in Europe they are used candied as a dessert and to some extent in the liquor industry as a flavoring.

Angelica's root can be eaten in its raw, whole form, by simply carrying a portion of the root and nibbling at it from time to time. Generally, the root is used as a tincture, thirty to sixty drops up to four times a day. The seeds can be tinctured also (ten to thirty drops up to four times a day), or several seeds can be taken in raw form and chewed.

Because angelica and osha are in the same family as hemlocks, they bear a slight similarity to some hemlock species. Osha can sometimes be confused with poison hemlock, angelica with water hemlock. The root and, to a lesser degree, the seeds and leaves, have a unique and readily identifiable smell, slightly celery-like but uniquely that of angelica. After picking, if you run your fingers along the cut root, its interior has a slippery, soaplike feel. The leaves are large, divided into smaller leaflets, and a lengthened oval in shape, perhaps three to four inches long. *Since angelica can be confused with water hemlock, which is quite poisonous, certain identification is a must.*

The plant often grows up to six feet tall with an impressive stalk topped with the characteristic umbel of flowers and seeds.[11] More than eighteen American Indian tribes used angelica species for medicine in a similar manner to that of Western medicinal use.[12] I have found it to be of particular use in normalizing menstruation and relieving cramping and nausea. *Its use during pregnancy is contraindicated.* The Creek Indians chewed the root and swallowed the juice or smoked it dry with tobacco for disorders of the stomach.

Coming upon angelica I am always struck by the feeling of femaleness and strong purity of spirit that the plant emanates. It is a shy plant, rarely in great abundance. It is a plant of water, always growing near a water source. In sitting in meditation with the spirit of angelica it is clear that the plant sits in balance between Heaven and Earth. The powerful hollow stem rises up and carries the spirit energy between the realms. Many shamans have carried the stem of angelica as a staff to help them maintain balance when traveling in spirit worlds.

The spirit of angelica is strong and may offer help to women who have an empty place within them (like angelica's stem). Go and sit with the plant and after making relationship with it ask (with the part of you that is empty) that it come into that place and reside there as an ally. Each plant carries a special spirit energy that can help when used as spirit medicine. Angelica is a plant of the east and it embodies those qualities. It carries within it vision, wisdom, enlightenment, and mature balance.

Its main strength and power is that of mature spiritual and emotional balance and strength in the face of conflicting demands from the environment, whether spiritual, emotional, or physical. I use this plant often in helping women who feel out of balance with themselves, suffer from hysterical asthma, whose reproductive organs are out of balance, who tend to lose themselves in others, suffer from anorexia nervosa, and for those who need a model of spiritual balance and strength. In using the plant as spirit medicine it is not enough to just ingest it as a medicine. One must sit in relation with the plant and have it agree to help, one must develop a family relationship with the plant until it becomes as a sister.

11
A SHORT COMPENDIUM OF PLANTS AND THEIR SACRED USES

Besides the perhaps fifteen plants with which I have very deep relationship, there are perhaps another thirty with which my relationship is fairly deep. The rest of the hundred and fifty plants I know and use are in varying degrees of depth. What follows is information about usage, habitat, appearance, and sacred uses of fourteen plants with which I have lesser degrees of relationship.

As I worked with the plants and got to know them better, I found that some plants must be picked at the same time as others when making a medicine. That, in fact, when the gestalt of a medicine is held in the mind for a particular person, each of the plants in that medicine are companion plants to the others, and all must be gathered at the same time for the medicine to be most effective. Three plants that fall in that category are licorice, echinacea, and red root. Together they make a powerful medicine, each enhancing the power of the other. When gathered together this power is even stronger and, in fact, in certain cultures it would be held that they must be gathered together. As you develop your knowledge of plant medicines you will find that some plants desire to be gathered together and you will then have the beginnings of your own knowledge of companion plants.

In this section I want to share a comparison with the sacred orientation in some indigenous cultures about a particular plant and contemporary medical uses. As such, this chapter does not go into the degree of detail as the last chapter.

I have included sacred songs, prayers, origin myths, or tales of usage of these plants from five tribes in North America: Hupa, Catawba, Sioux, Kwakiutl, and Ojibwa. The Catawba information used is the simplest, being an account in original language of how the plant is used. The Sioux and Ojibwa information contains the sacred song for the plant and sometimes an account of how it was received in vision. The information from the Kwakiutl tribe contains the usage of the plant, how it is prepared, the prayers said when harvesting it, and sometimes the prayers used when administering it. The Hupa information contains the sacred origin myth of the plant, the generic prayer used when harvesting it, and the intended usage. The Hupa stories all concern a sacred being that existed before people (the Indian) came to this world and who discovered a special healing use for a plant. The sacred being gave the knowledge of the healing power of the plant to the Indian when he came to exist on the Earth. I have tried to retain the style of the stories as they were originally told.

Yellow Dock
Rumex crispus

The leaves of the yellow dock are curly-edged, perhaps a foot in length, and alternate along the stem. The plant turns a striking rust-red in the fall; during the winter the dead stalks and heavy seed panicles at the top of the two-to-three-foot stem are readily recognizable. The root is shaped something like a carrot and is reddish brown outside, yellowish to orange inside. The darker the yellow of the root the stronger the medicinal qualities of the plant. The compounds making the root yellow are the primary medicinal substances of the plant and therefore are more present the more yellow the root. Plants growing in water are not acceptable, gathering should take place in drier embankments and open meadows.

The root is what is gathered for medicine. The primary uses of the plant are for indigestion, constipation, blood and skin disorders, and liver congestion. The root is effective in post-hepatitis flareups and difficulties in processing fatty foods. It has a mild but effective laxative action, much gentler than *Cascara sagrada* and other, stronger herbal laxatives. It is quite effective in combination with licorice for this problem.

I use the root extensively in formulas to treat psoriasis, combining it with sarsaparilla, dandelion root, burdock, Oregon grape root, and figwort in equal parts, one teaspoon three times a day. In general, the root can be thought of as a "blood cleanser," laxative, and treatment for liver congestion. To use the dry herb, two teaspoonsful of the ground root should be simmered for fifteen minutes in a cup of water and taken three times a day. The root tincture can be used effectively, up to thirty drops (the equivalent of one full dropper), three times a day.[1]

Kwakiutl Prayers and Uses of Yellow Dock

Among the Kwakiutl Indians when a woman has a swelling in her throat or has had indigestion for a long time she asks her husband to go and seek the roots and she tells him to pray. He goes into the woods and when he finds the plant he prays to it in this way:

> *I have come, I have reached you, Supernatural Ones, to come and*
> *pray of you to go and, please, heal my poor wife who for a long*
> *time has had a stomach ache, you, merciful ones, and please make*
> *her well, you Supernatural Ones.*

After he has dug the roots he takes them home and shows them to his wife and they both offer thanks. The husband washes the roots clean, places them in hot water and lets them boil a long time. When the roots are falling to pieces they are done. He pours the yellow dock into a small dish and gives it to his wife. She takes the roots and holds them, looking into the bowl, while praying.

> *Come, Supernatural One, I have got you by inviting you to come and, please, make me well, Supernatural One. Now you will go into me that you may drive away this sickness that troubles me, you pitiful one, Supernatural Ones.*

Then she eats the yellow dock and the juice. When she has finished the man takes the kettle and puts it back on the fire. He reheats the remaining yellow dock and liquid. When it is hot he takes some cotton cloth and pours the mixture into it. He folds it up and squeezes the liquid out of the mixture. Then he takes the hot cloth with the yellow dock in it and spreads it on the stomach of his wife above her navel. As soon as he puts it on her stomach he prays.

> *O Supernatural One, now help each other with your other part that has gone inside the stomach of our friend that, please, our friend may get well quickly, Supernatural One, Long-Life Maker.*[2]

Soon, the stomach of the woman begins to rumble. Then she defecates and the pain in her stomach is gone.

Wild American Licorice
Glycyrrhiza lepidota

Although there has been a great deal of discussion among herbalists (each quoting endless sources) regarding the comparison between wild American licorice and its relative, *Glycyrrhiza glabra*—the European licorice more commonly used in herbalism—I have found our American cousin to be as effective as the latter.

Wild American licorice is a member of the pea family and grows in large colonies formed and connected by creeping rootstalks. The roots are the medicinal part of the plant and they are easy to harvest, growing fairly shallowly beneath the soil. The American licorice is not sweet as is the European licorice. It tastes much more "pea-like." One of the benefits of the sweet variety of licorice is to make the rather strong taste of most herbal potions more palatable.

The leaves are made up of many smaller leaves along the leaf stalk (typical pea family) and are odd numbered, eleven to seventeen in number, with a single leaf at the tip. The plant stands one-and-a-half feet to three feet in height and has white to yellow-green, cloverlike blossoms. The seed pods of the plant are numerous and are a dark rusty brown and are burred, sticking easily to clothing. The burr-covered seed pods make this plant easily identifiable. The plant likes to grow along irrigation ditches and slow moving streams from 1,000 to 9,000 feet. It is often quite prolific when found, so harvesting an adequate supply of roots is not often a problem. The tap roots as opposed to the runners are stronger and if you wish to spend time attacking the often tough ground in which licorice grows, they may be the preferable harvest.

Licorice is one of the major herbs to know and a useful adjunct to any herbal repertory. Licorice is potent and has application in four major areas: lungs, adrenals, stomach, and the female reproductive system.

Within the lung system licorice provides several important helping actions. It is antispasmodic to reduce coughing, an expectorant to move phlegm up and out of the lungs, an anti-inflammatory to reduce lung inflammation suffered during onset of a cold, and demulcent to soothe and coat inflamed mucous passages.[3]

For people who have lit the candle from both ends, drinking huge amounts of coffee, constantly tense with an activated "flight or fight" response, and perhaps using cocaine or amphetamines, the usual result is exhausted adrenal glands. Since the adrenals are also the site of production for sex hormones, exhausted adrenals may lead to decreased libido. The adrenals also supply the substances that help facilitate healthy body repair and growth of bodily tissues. Exhausted adrenals can lead to longer healing time and a lack of vitality in body tissues. Licorice supplies many substances that can be used by the body as substitutes for those produced by the adrenals. Using licorice in conjunction with reducing stress and coffee or tea intake can allow the adrenals to rest and recuperate. Usually in cases of adrenal exhaustion licorice should be taken three times per day, 30 drops (one dropperful) per 150 pounds of weight for three to six months.[4]

Licorice also has strong application in treatment of ulceration in the stomach. The dried, powered herb should be used. If the ulceration is present in the stomach one ounce of the powdered herb should be added to any suitable liquid and drunk each morning. Ulceration can also occur in the duodenum, located just below the stomach in the first part of the intestines. The stomach and duodenum are separated by a sphincter muscle at the base of the stomach. For ulceration in the duodenum the powdered herbs should be taken in capsule form—two capsules, three times a day. When the capsules dissolve, the powdered herbs form a thick mass at the entrance to the duodenum. When the sphincter muscle opens the herbs fall, fairly intact, into the ulcerated area.[5]

Licorice is quite high in phytoestrogens, plant precursors to estrogen. I have found the herb to be very effective in the treatment of menopausal symptoms. Usually I combine it with other herbs high in plant estrogens such as hops and black cohosh. However, taken alone in sufficient dosage, it can be quite effective on its own. I suggest one dropperful six times a day to start, and the dose to be adjusted as needed.[6]

Licorice is antibacterial (against gram-negative bacteria) and antimicrobial. It also possesses antiviral properties and stimulates interferon production in the body. Licorice has shown effectiveness against bacteria that are resistant to standard antibiotics.[7] Further, licorice seems to enhance the action of other herbs taken in conjunction with it. Echinacea, when taken with licorice, produces immune-enhancing actions far beyond what it accomplishes on its own. The following is a very effective combination for colds or flu.

Herbal Flu Shot: Equal parts of tinctures of red root, echinacea, and licorice are combined. One dropperful of this mixture should be taken each hour at the onset of symptoms. It must be taken each hour to be effective.

Sioux Prayers and Uses of Licorice

Eagle Shield, the Lakota Sioux medicine man, was brought the knowledge of the use of licorice from visions he received from the bear. He was given a song during his vision, which he sang when administering the herb.

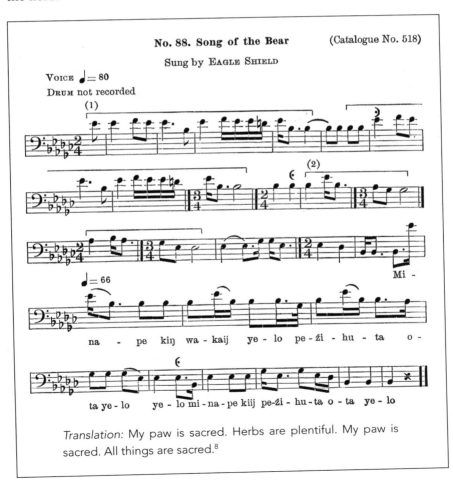

Translation: My paw is sacred. Herbs are plentiful. My paw is sacred. All things are sacred.[8]

Red Root
Ceanothus spp.

In California, red root grows often as small trees. In Colorado, the common species is *Ceanothus fendleri*. This type of red root is most often encountered as a scruffy, semi-thorny ground cover spreading over fairly good sized areas. The seed pods are a brilliant dark reddish color, about the same color of the tincture that is made from the roots. They are small, about half the size of a pea, and triangular. Before pods form, the branches grow small, fragrant clusters of white flowers, though in the California varieties the flowers may be lilac, pink, or purple. The stems have many sharp thorny projections and during harvesting can be painful. The exterior bark of the root is a dark to black color that, when scraped with the fingernail, will reveal a bright reddish inner bark. The more brilliant red this inner bark, the more potent the herb. The root is best harvested in the spring or fall when the reddish color is most pronounced. When most potent, the core of the root has a slight pinkish tinge. This tinge can be seen to extend throughout the whole root.

The root is very woody and should be cut when fresh into two-inch pieces. When completely dry the root is extremely hard and is very difficult to cut into smaller pieces. I usually collect the runners, which grow just beneath the surface. These runners make an imperceptible transition from root to stem and I mark this distinction where the black root bark begins to turn to the greenish color of the stem. Once I needed to place a post in a patch of red root and found the roots extending nearly four feet under the ground where they were the thickness of the handle of a baseball bat. They lasted a long time and made very good medicine.

Red root is a remarkable herb, primarily for the lymph system. It helps the lymph system process waste cells very quickly and reduces the time the body suffers from colds or flu. A number of herbalists have reported that red root increases T-cell count and is a useful adjunct for immune system disorders including AIDS. Red root is beneficial for inflamed tonsils, sore throat, and enlarged lymph nodes of any location. For inflamed tonsils or sore throat, the herbal tincture should be taken in the mouth, mixed with saliva and allowed to dribble directly down the throat onto the affected areas. Recommended dosage is 30 drops of the tincture per 150 pounds of weight, three times a day for chronic conditions, up to six times a day in acute episodes.[9] In spite of its large variety, all red root species are usable as long as the root bark and inner pith is reddish purple or brown-red in color.

Catawba Uses of Red Root

tu'ktu keri-keri 'here
wi'ti keri!'here hi-samu!' itci'gene tci-tci!samu'
isi'grere ata'de!. ta'ktu ti-ki!ye keri!howe i-ha'sigri
motu!' taktu'ki o'wetca ta'de mi-hari-yu'we
yi-tero ki mi-hari-ho'we.

Translation: Red root is good indeed. It is a good medicine root for the mouth of a child. When the little one's mouth is sore, wash with it. The roots of red root will prove good when nipples become sore. Using the red root make it up into medicine. You will be better and the little child will be better.[10]

Devil's Club
Oplopanax horridum

Devil's club is a member of the ginseng family and like most ginsengs it is helpful for the general vitality and health of human beings. The plant grows primarily in the northwest Americas and is a formidable plant. The stems grow from 6 inches to 12 feet in height and are usually an inch or less in diameter. They possess frightening needle-sharp thorn that literally cover most of the exterior of the stem. Imagine a cane pol with thousands of steel needles protruding from the shaft and you wi have a clear picture. The plant grows upright until its height become too much for the root and then it falls over, new roots forming along it length. The recumbent stem remains perhaps five inches above the fores floor. The plant prefers dark and moist old-growth forest where it is di ficult to see. Most people find devil's club for the first time when walk ing through such a forest: suddenly they find themselves tripping over recumbent stem and falling headlong into a patch of the herb, somethin like falling into a Stephen King vision of a demon blackberry patch.

Though I have successfully used the tinctured whole stems, mos harvesters use the bark of the root and maintain it is a much more pow erful medicine.

Devil's club moderates the symptoms of adult-onset diabetes an hypoglycemia. The herb smoothes out the high peak and valley expe rience of hypoglycemia and seems to keep the amount of sugar in th blood at a constant level. Extended use of the tincture in conjunction wit. American ginseng is quite useful for men over 55 who are lethargic, expe rience decreased libido, and get sick a lot. The herb can be used quite lik Siberian ginseng and will elevate the vitality levels in the body, eventuall restoring natural immune system levels, skin color and tone, and stamina Dosage: full dropper (30 drops) three times a day, six months to a year.[1]

Kwakiutl Prayers and Uses of Devil's Club

When a woman has laid long in bed with pains in her body she asks he husband to go and get the devil's club. So taking a mat and a small axe he goes along the river until he comes to the place where the devil's clu grows. Then he sits on the mat and he prays.

Now look at me, for I come trying to come to you, great Super-
natural Ones, being sent by my poor wife and I come and ask you

for mercy that you, please, take out her sickness, that is the reason
she has been lying in her bed for a long time sick with pains in her
body. Now I come to call you, Supernatural Ones, that you may
go, please, and save her, you, Life-Giver; and you also,
Healing-Woman, that you, please, may set right my poor wife,
please, she may live, please.

After he has prayed this way he takes the axe and he chops down four of the devil's club plants and then he chops off their tops. He then lays them on the mat he has brought with him, wraps them up and carries them home. When he gets there he lays down the mat in front of the fireplace and he makes up a fire in the fireplace. When it is going well he takes up the tongs and takes a piece of devil's club in the tongs and holds it in the fire until all the spines are scorched off. When they are all scorched off he takes up the devil's club and peels off the bark. When all the bark is removed he takes it and puts it into a kettle, fills it with water, and puts it on the fire. Then he boils it the whole day. When it is evening he takes it off the fire and carries it to the side of the bed where his wife lays sick. Then a tub he places there. With the tongs he takes out the bark and places it on the floor. Then he pours the liquid from the boiled devil's club into the tub. He places a board across the tub and his wife gets up and sits on the board. The husband places two blankets around his wife and lets them hang down around the tub to keep the steam from escaping. When she can stand the heat no longer she goes again and lies down in her bed.[12]

Seneca Snakeroot
Polygala senega

Seneca snakeroot grows in rocky woods from Georgia to Arkansas to North Dakota. It is a perennial six to eighteen inches high; leaves are alternate, lance shaped. The flowers grow in a terminal spike and are small, white, and pea-like. The medicinal part of the plant is the root.

The root is used primarily in lung conditions as an expectorant and antispasmodic. It is useful for colds and flu, for croup, pleurisy, chronic catarrh, asthma, and coughs. It is also used as an anti-inflammatory in rheumatism and as a poultice for swellings. The root also induces sweating and is a moderately good diuretic.[13]

Ojibwa Prayers and Uses of Seneca Snakeroot

The Ojibwa call Seneca snakeroot *bi'jikiwuk'*. The name translates, literally, as "buffalo medicine" but since the coming of white men and their cattle it is now referred to as "cattle herb medicine." When made into medicine among the Ojibwa, seneca snakeroot is usually combined with one to seven other herbs, the resulting combination also called *bi'jikiwuk*. Seneca snakeroot is considered to be the principal herb, without which the preparation would not be efficacious. According to Main'gans, an Ojibwa medicine man, seneca snakeroot came to the Ojibwa people as follows:

Pouch and Measure for Bi'jikiwuk'

There was once a *Mide'wini'ni* [a male member of the Mide'wiwin medicine society] who dreamed that he saw horned animals resembling [buffalo], under the water. They came up from the water and talked with him, telling him how to prepare this wonderful medicine. In order to persuade them to return he composed and sang a song. He was a young man at the time, but he sang this song until he was old. He sang it whenever he dug the roots or prepared the *bi'jikiwuk'*. Others learned it from him and now it is always sung when this medicine is prepared.[14]

Odjib'we, one of the most powerful of the Ojibwa medicine men, sang, for Frances Densmore, the song that had originally been composed in vision for use when digging and preparing seneca snakeroot.[15]

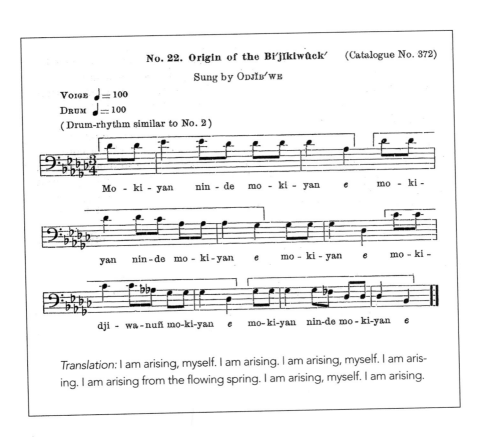

No. 22. Origin of the Bi'jikiwŭck' (Catalogue No. 372)

Sung by Odjĭb'we

Voice ♩ = 100
Drum ♩ = 100
(Drum-rhythm similar to No. 2)

Mo - ki - yan nin - de mo - ki - yan e mo - ki -

yan nin-de mo - ki-yan e mo - ki - yan e mo - ki -

dji - wa - nuñ mo-ki-yan e mo-ki-yan nin-de mo-ki-yan e

Translation: I am arising, myself. I am arising. I am arising, myself. I am arising. I am arising from the flowing spring. I am arising, myself. I am arising.

Bi'jikiwuk' was the principal war medicine carried by the Ojibwa. It was said to make men strong and to be a powerful healing medicine. An Ojibwa warrior's custom was to chew it and spray it from the lips on his body and equipment. The herb was considered to be effective in counteracting negative influences directed toward a person. It was taken four times daily all throughout life and was considered to enhance and increase the vitality and personal power of the person taking it.

Echinacea
Echinacea angustifolium

Echinacea has become perhaps the most used herbal medicine of our time, an "herb of the day." Echinacea is a "coneflower"-type plant growing throughout Arkansas, Texas, Montana, Wyoming, New Mexico, Kansas, and Nebraska, and occasionally in Colorado. It is perennial, growing from one-and-a-half feet to three feet in height. The flowers, daisylike, are light purple to light pink. The center of the flower is hard, raised, cone shaped, prickly brown. Though the roots are primarily used, the fresh seed heads are extremely effective medicine and should be prepared fresh. The roots can be dried but to maintain potency, should be kept whole until used.

Echinacea enhances the action of the immune system, increasing the number and response time of white blood cells and their ability to digest bacteria. The overuse of this herb for this purpose is regrettable for three reasons. First, "what goes up must come down." When the immune system is artificially stimulated for too long the unenhanced immune system deteriorates even further. Echinacea (if used for immune enhancement) should be taken only for short periods of time and should be disease specific. That is, if you are getting a cold, take echinacea (preferably in combination with other supportive herbs such as licorice and red root). If you are generally feeling run down and get sick a lot, don't take echinacea. Work on your immune system instead by taking Siberian ginseng for a year and change your diet and eliminate as much stress as you can. Second, the extreme popularity of echinacea has resulted in wholesale environmental degradation in some areas as large companies use backhoes to dig up miles of prairie to get at the plant. Third, the herb is effective only if taken for a maximum of ten days straight. After that its usefulness diminishes rapidly. If for some reason you do need to take it for extended periods (for immune support), you should cease intake after ten days, rest a couple of days, then begin again.

Echinacea is also extremely useful in speeding tissue repair and in healing connective tissue. Torn ligaments respond well to echinacea and it is also of great use in healing salves for tissue repair. In combination with St. John's wort as a salve, scarring from surgery and other wounds can be reduced markedly.

Echinacea is also of tremendous use in strep infections, particularly strep throat. The herb numbs affected throat tissue which helps alleviate symptoms but it is also directly effective against the infection itself. The herb should be taken as a tincture and taken by mouth. Mix it with saliva and then allow it to dribble over the affected area. A full dropper should be taken in this manner each hour until the disease is brought under control.

Though a useful and powerful herb, echinacea can be used incorrectly. It is not a panacea. Conscious intention and awareness of its strengths and limits are essential for the best outcome.[16]

———————— 🌿 ————————

Sioux Prayer and Uses of Echinacea

Bear-with-White-Paw, a Lakota healer, used echinacea as one of his principal medicines. He used it for tonsillitis, pain in the bowels, and toothache. Before using the medicine he would say the following prayer:

Ho mita'kola on masi'ke tanyan' ana'magoptan yo. Anpe'tu kin lehanl'
ta'ku wanzi'awa'can mi he oci'ciyakin ktelo'. Pezu'ta lena'ke slolye'
mayaki'yin na ta'ku econ' maya'si na wo'yazan wan'zigzi api'ye ma'yasi

na hena' iyo'kihe kta ke'he kon wana' lehanl' iya'lipe wa'yin kta tka he'ce
iyu'ha owa'kihi kta. Wy'yazan hena' heyab' iya'yin kte.

Translation: My friend, I am poor and needy. Listen well to me. This day I have something in my mind, and I wish to tell you. All these medicines you have made known to me, and you have commanded me to perform certain things in order to attend to certain sicknesses, and you have told me that these medicines have certain powers in them. Now I wish to use them with effect. These sicknesses, I want them to go away.

After he said this he would sing the following song when using the herb:[17]

No. 90. Song of Healing (Catalogue No. 674)

Sung by BEAR-WITH-WHITE-PAW

VOICE ♩ = 66

DRUM not recorded

Translation: Father, come forth! A two-legged object (i.e., a man), lying in the earth, I have renewed. Father, come forth! Mother, come forth! A two-legged object, lying in the earth, I have renewed. Mother, come forth!

Bear-with-White-Paw

Redwood Sorrel
Oxalis oregana

Redwood sorrel is used in the same way that many of the sorrel plants are used. The above ground plant is used medicinally. The herb has traditionally been used for scurvy, scrofula, fevers, and skin diseases and as a poultice for tumors and boils. The leaves have a rather sour, acidic taste like most of the sorrels and are usually used fresh. They are chewed for mouth sores, nausea, and sore throats. When used for a poultice the herb also should be fresh. For scurvy the fresh plant or leaf tea is used. The fresh leaves are also good in salads or as a pot herb. Large doses can cause oxalate poisoning; the proper dose is 2–4 cups of the leaf tea per day.[18]

Hupa Origin Story and Use of Redwood Sorrel

Formula of Medicine for Pregnant Women (Bear's Medicine)

nin-nis-an min-nĕ-djit hai-ya a-na-tcit-yau sats na-wa-
World in the middle there she got that way bear while

2 nĕ-djōx kyū-wiñ-ya-in-yan xon-nis-te me teL-tcwen hai-ya
walking. Offspring her body in grew. There

e-il-wil ke-ûL-ᵃx La xû-Le ke-ûL-ᵃx min-nĕ-djō-xō-miL dō-ma-
all day she every night she chewed. After a while
chewed,

4 a-din-niL-tcwit hai-ûñ wûñ-xoi-kyûñ-na-ī-ya da-xwed-dûk-kī-
she couldn't walk. Then she began to think about it, " Why

auw auw-dī-yau a-dim-mit ta-xûñ-hwō-ōw mit-tis na-teñ-en
am I this Her own so large over it she looked.
way?" belly

6 kût auw-xa-tin-ne kyū-wiñ-ya-in-yan-ta-diñ hai-ded auw-dī-
" I wonder if this Indian world this way I am."
way it will be

yau hai-ya-miL xwen-neûk-kût-tciñ tce-xa-neûw, a-den-ne
And at her back it spoke. It said,

8 hwe dōñ sa-hwil-lūw hai eñ dōñ kyū-wiñ-ya-in-yan ma
" Me put in your " It is Indians for
mouth."

a-nûñ-dī-yau hai-ya-miL deōx na-teñ-iñ-hit ded ûñ-kya La
you are that And around when she looked this she saw, one
way."

10 xau-dik kit-tûn-dûñ-qōtc† hai-ya-miL sa-wil-lai yis-xûñ-hit
standing its leaves sour. And she put it Next day
in her mouth.

ûñ-kya na-na-is-ya-e-xō-lan hai-ya-miL a-tcon-des-ne hai-yûk
she found she could walk. And she thought, "This way

12 xō-lûn-teL kyū-wiñ-ya-in-yan-ta-diñ hai-ded kī-ma-ū miL hai
it will be Indian world, this medicine with. This

kût hwit-Lō-we hai-ye-he dō-xō-liñ Lan hwin-nis-te tcō-
is my medicine. Anyhow not many my body

naL-tsit-te kyū-wiñ-ya-in-yan-ta-tciñ na-de-tcwit-te hai miL
will know. Indian world in I will leave it. It with

a-hwiL-tcit-den-te
they will talk to me."

hai-yūx tcin a-na-tcit-yau
This they say she did.

Translation: While walking in the middle of the world, Bear got this way. Young grew in her body. All day and all night she fed. After a while she got so big she could not walk. Then she began to consider why she was in this condition. "I wonder if they will be the way I am, in the Indian world?" [Then] she heard someone talking behind her. It said, "Put me in your mouth. You are in this condition for the sake of Indians."

When she looked around she saw a single plant of redwood sorrel standing there. She put it into her mouth. The next day she found she was able to walk. She thought, "It will be this way in the Indian world with this medicine. This will be my medicine. At best not many will know about me. I will leave it in the Indian world. They will talk to me with it."

This they say she did.

Hupa Prayer before Picking the Medicine

10 hai-ded   eñ   min-Lûñ   mit-da-Lit-yōs   miL   na-ō-dūw-de-xût
    "Here   are   ten   packages.   With   I ask you for it.
                                                   them

hai   nit-Lō-we   hwū-wa-meL-tcwit-te   hwe   kûn-nañ   dō-hwa-da-
       Your   lend me.   I   too   I have not
       medicine

12 da-tsin   kyū-wiñ-ya-in-yan-ta-diñ   til-tcwen   hai-ya-hit-djit hei-
    got much   in Indian world   grows."   And then

    yûñ   kût   xōw-dōñ   kût   hwit-dje   ōL-tsit   kûn   nū-wa-me-neL-
    "Yes   I guess it   kût   My mind   you know.   I will lend it
            is so.

14 tcwit-te   yō   dōñ   kis-xûñ   hai   hwit-Lō-we   mī-ye   nōñ-xauw-ne
    to you. There   it stands   that   my medicine   Under   put it."

    hai-ya-hit-djit   xe-e-iL-yōl   hai-ded   mûx-xa-tce-xō-len
    And then   he blows   that   incense.
                       away

Translation: To the plant one brings an offering of tobacco and incense root. And offering it up to creator one says, "Here are ten packages of tobacco. With them I ask you for the herb. Lend me your medicine. I do not own much of the tobacco that grows in the Indian world."

And then Creator responds: "Yes, it is true I guess that you know my mind. I will lend the herb to you. There it stands, my medicine. Under the child put that medicine."

And then the offering of incense root is blown from the hand.[19]

Milkweed
Asclepias spp.

There are three main species of milkweed used in herbal medicine: the common milkweed, *Asclepias speciosa;* inmortal, *Asclepias asperula;* and pleurisy root, *Asclepias tuberosa.* While each are specific they also have certain similarities in their effects on the body.

The common milkweed, *Asclepias speciosa,* grows throughout the west. It is scattered and erratic but when found often grows in largish colonies. Inmortal grows primarily throughout the desert southwest and extends into western Nebraska and Arkansas. Pleurisy root, the most common milkweed of herbal use, is primarily an herb of the eastern United States.

Common milkweed grows from one-and-one-half to six feet in height, usually with a single stem. The leaves are broad, oval, and longish, growing up to eight or ten inches in length. The veining in the leaves is markedly pronounced. The plant produces flowers in a large globular cluster, pinkish white in color. The individual flowers are quite delicate and pretty and are very sweet. They may be cooked slowly in water and, when the liquid is boiled down, make a thick syrup suitable for pancakes. After flowering, the plant produces a good sized pod, three to five inches in length, filled with downy seeds.

Milkweed is a highly useful plant. The shoots may be gathered when four to eight inches in height and prepared like asparagus. The thick, white, milky sap contains a number of cardiac glycosides and is quite bitter. The water in which the milkweed is cooked should be changed one or two times. The sap, which is water soluble, is discarded along with the water. When so prepared the plant tastes much like asparagus. I feel it is even more palatable. Also edible are the leaves, which are much like spinach when prepared. The flower buds can be boiled as well and taste like a unique cross between peas and asparagus. The pods, before becoming filled with the downy seeds, are very good also, somewhat like okra and quite delicious.

The root of the plant, which is used for medicine, is often about the size of a pencil. It may continue down for some distance into the soil or just peter out with no warning. Occasionally the roots form thick clusters of tubers.

The root, when taken as medicine, is primarily an herb for the lungs. It softens bronchial mucous, facilitates expectoration, and dilates the bronchial passageways. Like its cousin inmortal, the herb stimulates lymph drainage from the lungs though not so markedly. As such it can find use in treatment of asthma, symptoms of colds and flu that are seated in the lungs, bronchitis, pleurisy, and chronic problems such as emphysema and cystic fibrosis.

The herb also acts well as a diuretic and increases sweating. Too much can cause nausea and, eventually, vomiting if one persists. I generally recommend to start with 15 drops of the tincture three to four times a day and increase dosage until the edge of nausea is just starting to make itself apparent, then back off a bit.

Pleurisy, an inflammation and painful drying of the sack that contains the lungs, was often treated with pleurisy root, which has a specific effect as an anti-inflammatory for the pleura. Most milkweeds have this effect, though in lesser strengths.

The narrow-leaved milkweeds possess larger quantities of the constituents that cause nausea and vomiting and should be used more cautiously. Milkweed sap is considered to be a remedy for warts when the milky juice is rubbed on the wart daily.[20]

----------------------------- ❧ -----------------------------

Catawba Uses of Milkweed

Milkweed Medicine for Snake-bite. (III)

sərá'k	witá'sɔre	ya'tcukʻha	wį'ti·	kərį'hɔre	ya	wį·ti·.
Flower	milk	snake bite	medicine	good is	snake	medicine.

wi·ta's	kį·	owatcaᵇ	yę'pa'	tcuk	hi·k	wa'we	mi'bari	huwe.
Milk	the	using	person's foot	bite	kills	may	get better	may.

ya'kį·		tcukʻdawa'riwe.		ya	sigri·hɔreᵛ.
Snake the	bite	somewhere go die will.		Snake	poisonous is.

Translation: The milk of the milkweed is good medicine for snakebite. It is good snake medicine. Rub it on the foot where the snakebite is. The snake is very poisonous. They may get better however; the milk is good medicine for snakebite.[21]

Oak

Quercus spp.

Oak is a widely available species throughout the United States, varying from small scraggly bushes to the great tree of legend. Because of the considerable variation in leaf shape, the only reliable method of identification is the fruit. Acorns, or their caps, will almost always be present on the oak or scattered at its base.

Although I tend to use wild geranium, or cranesbill as it is also known, more than oak in making medicine, oak is considered to be the traditional astringent in herbal medicine. Astringents contract or draw together organic tissues aiding in cell wall binding. This is particularly useful in wounds or abrasions to the skin. The tannin that oaks contain is the origin of their astringent action. The bark is generally used though the leaves and twigs can be used reliably. Oak galls are particularly potent medicine as they may contain up to two or three times the tannin content of the bark. Galls, sometimes call gallnuts, are a localized, generally round, ball-like growth on plants, caused by irritation at the site of its occurrence.

Oak is often used as a strong infusion or tea for gum inflammations, skin abrasions, sunburn, bleeding wounds, diarrhea, dysentery, and hemorrhoids. One teaspoon of the bark powder in a cup of water is boiled gently for fifteen minutes and applied to the affected area. For bleeding wounds, I have found the powdered herb (leaves, twigs, bark, or galls) to be especially effective when applied directly to the affected area. Often I will use the oak in combination with powdered antibiotic herbs such as usnea or echinacea. Gum inflammations respond well to a tincture of the bark, other complaints respond well to the strong infusion or tea. For use in diarrhea or dysentery drink the tea, prepared as above, three to six cups a day. These conditions also respond extremely well to the use of the tea as an enema.[22] Oak enemas were, at one time, a primary approach to the treatment of extreme cases of dysentery.

Alleviation of the symptoms of dysentery through the use of oak enemas is quite rapid and seems almost magical.

Hupa Origins and Uses of Oak

Formula of Medicine for the Birth of the First Child

Yimantuwinyai lived at Teexoltewedin. But one day, down river he went. To TseLtcedin he went. Going along the beach, going south. Then he went along the ridges. From the north, south along the ridges he went. And then he sat down. And from the south he noticed two women coming.

"Rest you, too," he said and they sat down.

They sat down and looking at them, he began to think, "I would like to have intercourse with some woman."

Then he left that place and the women went on toward the north.

And then he, himself, went south, walking along. That night he spent at this place south of Willow Creek. He found a big rock and there was this place under it and he spent the night there. There again two women came and spent the night. Him under the rock, they above.

And he thought again, "I would like to have intercourse with some woman."

Then in the morning, they went north and he, himself, went south. He went south along the middle of the world, walking along until he did not feel like going further. He rested. Again he began thinking about women; he did not feel like going on without having intercourse with some woman. Then he was surprised to see a white oak growing there and there was this hollow place so he had intercourse with that oak. And then he went on again, going south, coming finally to the world's end.

Then, when he was ready, he left that place and started back, going north. From the south he came back, until he came again to that place where the white oak grew. And as he came along there he began to hear a baby cry and this caused him to remember how he had had intercourse with that tree going south.

He came up to the tree and stood facing it. He was surprised to hear that baby's cry coming from inside it. Inside that hollow tree he heard it cry. How is this? he thought.

Around the hollow place a knurl had grown and so he pounded it off. He took the bark he had knocked off and he made a tea with it, then on that place he poured it. He poured it on there and it really opened up that white oak.

"This way it will be," he thought. "When the Indians come. This way it will be easy for them."

And that way he took out the baby.

Before gathering any plant medicine the Hupa sing the "Prayer before Picking the Medicine." See page 157 for the full text and translation of this prayer.

Yarrow
Achillea millefolium

Yarrow grows the world over and this perhaps explains it popularity among indigenous tribes. It is one of the most widely recognized and used herbs among all indigenous peoples. The plant forms a mat of basal leaves, closely fernlike, in the spring. It grows upward in a single stalk, eight to twelve inches tall, and produces a flat cluster of white flowers around July, depending on the altitude. The leaves when crushed have a pleasant, almost minty smell, unique to yarrow.

Though some people prepare only the flowers as medicine, I commonly use the whole plant but leave the root. The root may be used with effectiveness but I prefer to leave it so that it can continue to produce new plants each year.

Prepared as a hot tea, yarrow will stimulate perspiration. I have found the tincture useful for settling the stomach, though it is more effective in combination with other herbs such as betony and poleo mint. The herb is traditionally used to aid the body in diseases where fevers are present. It lowers the blood pressure, though marginally. I have found it effective used in conjunction with passion flower for this purpose. It stimulates digestion and tones the stomach. As such it is useful in cases of hiatal hernia where the stomach has lost tone and is protruding up through the esophagus or a tear in the muscle wall. The fresh leaves are an excellent remedy for bleeding; placed on moderate sized to small cuts they readily stimulate clotting. The tea or tincture is also excellent for decreasing menstruation during a particularly strong flow—use a full dropper of tincture every three hours.[23]

Sioux Prayers and Uses of Yarrow

The Sioux healer, Eagle Shield, kept yarrow in a special bag for use with patients. He used the herb for those shot through the body. Eagle Shield called it *tao'pi pezu'ta*, herb for the wounded. The whole plant was collected and dried and the patient was required to chew it.

He sang the following songs when treating people with yarrow. Between renditions of these songs he gave deep groans, like those of a man in extreme distress, frequently interspersing these with a hissing exhalation of the breath.

No. 81. "Behold All These Things" (Catalogue No. 511)

Sung by EAGLE SHIELD

VOICE ♩ = 66

DRUM not recorded

I - ho le - na waij-yaij- ka yo i - ho le - na waij-yaij- ka yo

ta - ku waij he - ħa - ka - se waij - la - ke ɓiij

on ya - ni pi kte - lo

Translation: Behold all these things, something elklike you will behold. You will live. Behold all these things. Something buffalolike you will behold. You will live.

After the patient began to improve, Eagle Shield sang a second song with the use of yarrow.

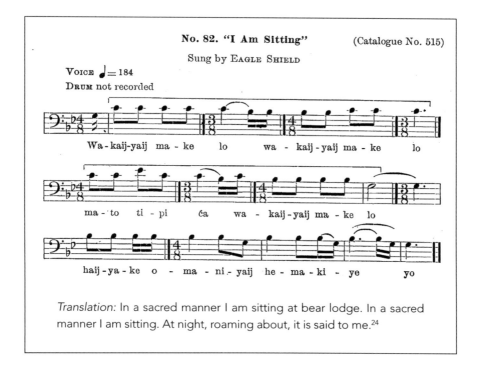

No. 82. "I Am Sitting" (Catalogue No. 515)

Sung by EAGLE SHIELD

Voice ♩ = 184
DRUM not recorded

Wa - kaij-yaij ma - ke lo wa - kaij - yaij ma - ke lo

ma - to ti - pi ca wa - kaij - yaij ma - ke lo

haij - ya - ke o - ma - ni - yaij he - ma - ki - ye yo

Translation: In a sacred manner I am sitting at bear lodge. In a sacred manner I am sitting. At night, roaming about, it is said to me.[24]

Alder

Alnus spp.

Alder is a shrub or tree that grows up to fifteen feet in height. The leaves are wedge shaped, slightly rounded at the base and wider across the middle. It grows throughout North America. It forms thickets along waterways. In Colorado the primary species is *Alnus tenuifolia* and is common in the foothills to subalpine areas. The tree flowers in the spring and berries ripen in autumn. Alder is used almost identically to oak. The bark is high in tannin and can be used in all situations where tannin is effective such as: (internally) diarrhea, gum inflammations, and sore throats; and (externally) as a wash for cuts, hives, poison ivy, swellings and sprains. The bark tea has a shrinking, clotting, and antiseptic effect and is therefore good for wounds. The bark is best collected in the spring or fall, though is good anytime. The bark is best used fresh or of recent collection. Make a strong decoction for internal use, weak for external.[25]

———————— ✦ ————————

Kwakiutl Prayers and Uses of Alder

When a woman is spitting blood she asks her husband to go for herbs for her. He takes his axe and canoe and goes to a place on the river where an alder patch is growing. When he finds it he gets out of the canoe and sits down under the alder he has chosen. Then he looks at it and prays.

> *I have come to ask you to take mercy, Supernatural-Power-of-the-River-Bank, that you may, please, make well with your healing power my poor wife who is spitting blood. Go on, please, pity me for I am troubled and, please, make her well, you, Healing-Woman, you, Supernatural-Power-of-the-River-Bank, and, please, stop up the source of blood, you Causing-to-Heal-Woman, and, please, heal up the cause of trouble of my poor wife, please, you, great Supernatural One.*

Then he stands up and with his axe he chops off four pieces of alder bark and goes home. When he arrives he cuts off the rough outer bark of the alder and cuts into smaller pieces, bite size, the alder bark. The woman holds the pieces of alder bark in her hand, and looking at them, she prays.

> *Welcome, Supernatural One, you have come to take pity on me that I, please, may get well. Go on, please, try with all your strength, you, Well-Walking-Woman and, please, heal me, you Healing-Woman and, please, save me, you, Life-Maker and, please, hit this sickness that it may die sooner than I, please.*

When she has said this she puts one of the cut pieces of alder bark into her mouth, chewing it and making her saliva full with its juice. Then she sucks it dry and swallows the juice. This she does with all the pieces of alder.[26]

Wormwood
Artemesia spp.

Many species of wormwood, also known as mugwort, are called "sage." This is a misnomer as they are not related to the sage family and cannot be used interchangeably. In fact, many of the wormwoods have distinctly different actions when used medicinally. True sages decrease secretions in the body. They can be used effectively for decreasing mucous production, lactation, sweating, and menstruation. Wormwood species, however, can stimulate some secretions in the body. They increase sweating and menstruation. Because wormwoods increase blood flow in the uterus, they should not be used during pregnancy. The wormwoods grow from sea level to 11,000 feet in altitude over much of the world. The usable plants are distinctly hairy and have a fresh, strong, pleasant scent when crushed. The flowers are small ball-like clusters, often yellow in color. The plant can vary from eight inches in height up to three feet in some California mugwort species or even taller for some big sagebrush species.

The root of the wormwood is rarely used. The above-ground leaves, flowers, seeds, and stems are harvested for use and can be picked at any time of the year. A rounded teaspoon of the dried plant in a cup of hot water, allowed to steep for fifteen minutes, is useful to promote sweating in feverish states or to increase scanty menstruation.[27] I have also used a full dropper of the tinctured herb, three or four times a day, for use in increasing menstrual flow.

Wormwood tea, though bitter, is a good remedy for stomach indigestion. It has also been traditionally used, as its name suggests, in cases of intestinal worms. I have never used it in this manner, though Michael Moore suggests two cups a day for at least two weeks, making sure its use is constant.

The herb has traditionally been used as a smudging agent. The green plant is cut and gathered together in a bundle and wrapped with small string and allowed to dry. The end is lit and used as a "smudge wand" in ceremonial smudging. The plant is also used, dry or moistened, in the sweat lodge. The plant is placed on the hot stones in the center of the lodge and the resulting vapor inhaled. I have used this herb with good success in the lodge for stimulating sweating. Additionally, I feel that the herb has the tendency, like sage and cedar, to dispel negative feelings or energy. I will often pick a few leaves when wildcrafting and crush them between my fingers, sniffing the resulting strong smell througout the day. I find it revitalizing and very stimulating.

Sioux Prayers and Uses of Wormwood

For those suffering from headache and lack of appetite, Eagle Shield would use wormwood. In the Lakota language the herb is called *nasu'la yazan'pi ipi'ya*, "no appetite medicine." The root was dried and powdered and sprinkled on hot coals and inhaled. Eagle Shield received a song in vision from the bear for use with this plant. Before singing it he would say the following prayer:

Peźu'ta ćiću' ktelo' tka waśto' ća yam' ktelo' lona'ko waśto' ke'yapo.

Translation: Herbs I shall give you, but (they are) good, so you shall recover, all these (are) good, they say.

At the end of this prayer he would sing the song he received in vision.

No. 84. "These Are Good" (Catalogue No. 513)
Sung by Eagle Shield

Voice ♩ = 80
Drum not recorded

Le - na - ke waś - te ke - ya - pe - lɔ ho - ye-
ya na-žiŋ

Translation: All these are good they say. With a noise they stand erect.[28]

Wild Ginger
Asarum canadense

A creeping perennial, the leaves of wild ginger are strongly heart shaped. The plant grows low to the ground and the maroon, urn shaped flower, growing between two leaf stalks, is often covered by last year's leaves. It is a beautiful little plant, liking rich woods, and coming up sometime between April and May. The root is round and fleshy with a strongly aromatic odor, reminiscent of the ginger of commerce.

The root is the part of the plant generally used medicinally. It should be harvested in fall or spring and used as medicine either fresh or dried. Both wild and commercial ginger are powerful stimulants for the circulation of the body. They promote sweating, menstruation, lactation, and circulation of the blood. I will often combine ginger with prickly ash bark and cayenne, though it is useful alone in helping people who experience extremely cold hands or feet in winter months. It is not uncommon for women who have chronically cold hands or feet to also experience ongoing menstrual problems. Flow may be scanty accompanied by severe cramping. I have also observed the presence of migraine headaches in women with those symptoms. Stimulation of blood circulation with the use of ginger can alleviate these problems.

Wild ginger is considered specific for painful cramping of the bowels and stomach. I will often use other herbs, however, such as feverfew for migraine and black cohosh for cramping. I use the herb primarily as a tincture, one-third dropperful three or four times a day. Because of its stimulation of blood circulation in the uterus, it may seem that the gingers should not be used during pregnancy. Commercial ginger, *(Zingiber officinale)*, however, has been used extensively for morning sickness and nausea and no side effects have been noted. Wild ginger has reportedly been used for nausea in folk medicine, and the effectiveness of commercial ginger makes this use fruitful to explore. Commercial ginger has also been found extremely useful in arthritic joint inflammation and any kind of nausea from motion sicknesses to stomach flu. Most cases reported that ginger taken in these instances was dried and powdered in capsules. Though some people have reported taking ginger in very large doses (500 mg capsules to twenty times a day) most trials reported the use of 250 mg or 500 mg capsules of the powdered root to four times per day.[29]

The tea may also be used, a rounded teaspoon to a cup of water, instead of the tincture.

Hupa Origins, Prayers, and Uses of Wild Ginger

Formula of Medicine to Cause the Infant
to Grow Fast and Healthy

Out East in the corner of the world, a maiden lived all alone. She saw nobody, not even the tracks of people did she see. But after a time, without reason, she became pregnant. And yet she had seen nobody.

And so she thought, "Where is it from that this baby has come?"

Eventually it was time for her to begin birthing that baby. And so she did, from her that baby came.

She thought to herself, "I will pick up that baby."

But it dodged away from her. She kept trying to pick it up but always it dodged. Finally, it tumbled down from the sky, and toward the west it went, getting so close it seemed about to tumble into the water. But close to the beach it stopped and where it stopped the plant grew. At the base of the plant it stopped, and when it did that the herb became medicine. Then the woman came to the baby and broke off the medicine. With it she picked him up. Then, back to the corner of the world, in the east, she took him. She took the medicine and steamed the baby with it and the baby grew fast and healthy.

Before gathering any plant medicine the Hupa sing the "Prayer before Picking the Medicine." See page 157 for the full text and translation of this prayer.

Elder
Sambucus spp.

Elder is an indigenous shrub growing throughout all regions of the United States. The plant is almost treelike, in many instances growing up to twelve feet in height. Elder plants often contain a profusion of woody stems coming out of a central root, much like some of the sumacs. The branching stems are often covered with a rough bark, the inner, larger stems are smooth. The leaves are ovate-oval to oblong-lanceolate and serrated, the flowers star shaped, fragrant and growing in a flattish cluster. The flowers give rise to hanging clusters of berries in late summer and fall. The berries are a darkish purple, never red, becoming nearly black when fully ripe. They are quite tasty when cooked and are used in jams, jellies, pies, pancakes, and wine. The flowers are also considered edible and are sometimes cooked into pancakes and fritters.

Folk legend has it that the continued use of elder will heal all the ills a person is likely to have in a lifetime, thus promoting long life and giving rise to its name, *elder.*

The inner, green bark and the leaves, as an expressed juice in doses of one-half to one fluid ounce, is considered reliably effective as an emetic. Lower doses will encourage gastric and fluid secretions. Because the bark is so strong, it is advisable to become informed in the use of the plant before using it too freely. Generally, the flowers and berries are most often used in conventional herbal practice. The flowers are excellent for upper respiratory inflammations such as colds, flu, and hay fever. The flower tincture (one dropperful to three times a day) will help clear mucous conditions in the upper respiratory tract, reduce inflammation, and help healing. The flowers, when steeped as a hot tea (cup of hot water to two teaspoons of the fresh or dried flowers), are considered to be a general stimulant for the body. When prepared in cold water (same proportions) they are felt to be a good laxative and effective diuretic. The berries are also often used for their laxative properties, primarily through taking a glass of the expressed juice, diluted with hot water twice a day. The flowers and leaves are often used in salves for wounds to soften the skin and help in general healing.[30]

Kwakiutl Prayers and Uses of Elder

When a woman is ill and cannot vomit her husband goes for elder root in the forest. The man gets his digging stick and finds an elder patch. He seeks a small elder and sits down in front of it. Then looking at the elder, he says:

> 0, Supernatural One, you who are not ordinary, I come to pray
> to you that you go, please, and make vomit my poor wife that she
> may vomit the cause which makes her feel qualmish all the time,
> you, Merciful-One, that you, please make come out the
> cause of it, Supernatural One.

As soon as he prays, he digs the root and cuts off a length the equal of four fingers, the thickness of one finger. As soon as he gets home he pours water into a small dish. Then he washes off the dirt from the roots. He pours out the dirty water and pours fresh water into the dish. He takes a rough stone, lays it in the dish and rubs the wet roots on it. When the water becomes milky, he stops. He pours the medicine into a cup and gives it to his wife. As his wife holds the cup, looking into it, she prays:

> The-One-Asked-to-heal, Supernatural-One, you have come to take
>   pity on me, you who have the name Healing-Woman, you who
> have the name Life-Bringer. Please, press out this which causes my
> qualmishness that I, please, may vomit, please, Supernatural One.

Then she drinks it. Sometimes a second cup of the medicine is prepared and more prayers said. This is all that is needed.[31]

12
THE BIRTH OF GAIA

[T]he relationship to the world that modern science fostered and shaped now appears to have exhausted its potential. It is increasingly clear that, strangely, the relationship is missing something. It fails to connect with the most intrinsic nature of reality, and with natural human experience. It is now more of a source of disintegration and doubt than a source of integration and meaning. It produces what amounts to a state of schizophrenia: Man as an observer is becoming alienated from himself as a being. Classical modern science described only the surface of things, a single dimension of reality. And the more dogmatically science treated it as the only dimension, as the very essence of reality, the more misleading it became. Today, for instance, we may know immeasurably more about the universe than our ancestors did, and yet, it increasingly seems they knew something more essential about it than we do, something that escapes us. The same thing is true of nature and of ourselves. The more thoroughly all our organs and their functions, their internal structure and the biochemical reactions that take place within them are described, the more we seem

175

to fail to grasp the spirit, purpose and meaning of the
system that they create together and that we experience
as our unique self.[1]

—VACLAV HAVEL
PRESIDENT OF THE CZECH REPUBLIC

*M*y journey with sacred plant medicine resulted in my connecting with a deeper world than the one in which I grew up. As I have worked with my sense of this over the years, I have come to understand that sacred plant medicine comes out of an even deeper reality. That reality is encoded in the Native American concept of Mother Earth and conveys in words the sense that the Earth is a single living being and the life upon it her children. This awareness seems to be prevalent in nearly all indigenous societies. But for Western man, scientific man, the concept is relatively new, reaching popularity only with maverick NASA scientist James Lovelock's work, which he has called the Gaia Hypothesis. Lovelock, examining the Earth ecosystem, noticed it was self regulating, and began to think of it as a single physiological system. As Lovelock says:

> This top-down view of our Earth as a single system, one that I call Gaia, is essentially physiological. It is concerned with the working of the whole system, not with the separated parts of a planet divided arbitrarily into the biosphere, the atmosphere, the lithosphere, and the hydrosphere. These are not real divisions of the Earth, they are spheres of influence inhabited by academic scientists.[2]

Lovelock goes on to note that any approach to living on Earth as a species cannot help but fail if this fact is not taken into account. This failure can be seen in many places. And the deep understanding that indigenous cultures possessed about Gaia is the more startling in contrast.

There is an old Navajo warning that, if you kill off the prairie dogs there will be no one to cry for rain. Bill Mollison, the father of Permaculture, notes in response that:

> Amused scientists, knowing that there was no conceivable relation-ship between prairie dogs and rain, recommended the extermina-tion of all burrowing animals in some desert areas planted to range-lands in the 1950s . . . in order to protect the roots of the sparse desert grasses. Today the area (not far from Chilchinbito, Arizona) has become a virtual wasteland. Fierce runoff, soil compaction, and lack of fresh seedbed have carried the grasses away.[3]

In fact, the burrowing animals like prairie dogs open breathing tubes in the Earth. The underground aquifers act like the diaphragm in human bodies, the moon as it passes raises and lowers the underground water table and the Earth breathes through the many fissures and tubes opened by the burrowing creatures. The exhalation of moisture-laden air, filled with negative ions, helps create rain.

In addition to this breathing mechanism of the Earth, the trees are also a major factor in the creation of rain. Trees account for half or even more of all moisture returned to the air. The forest and the ecosystem it creates accounts for most of the moisture in any bioregion. Cutting the trees results, in one person's lifetime, in the creation of drought. The tree, like most wild things, consists of all its interrelationships in the biosystem. The real identity of the tree cannot be found in its parts. Just like the ecosystem it is not a conglomerate of pieces. It is, rather, expressions of the same thing in different form. The tree does not end at its skin but exists also in the rain that falls downwind, many miles from the forest. In the seed exists the acorn, the oak, and the shade. The tree is also the fungi that live at its base, the bird that lives on its seed. As Mollison notes:

> [T]he tree can be understood only as its total entity which, like our-selves, reaches out into all things. Animals are like messengers of

the tree, and trees the gardens of animals. Life depends on life. All forces, all elements, all life forms are the biomass of the tree.[4]

And as Brooke Medicine Eagle notes in the foreword: "My elders have said to me: 'The trees are the teachers of the law.'"

I have often felt that the tree holds some special significance for human beings. It is the one plant that has a long tradition of a separate sacred archetype that is found in all religions. When people come upon an ancient tree, some event occurs that takes them out of themselves and they, for a moment, feel the touch of a greater and much deeper reality.

The destruction of the rain forests has elevated awareness of the crucial element the trees and forest systems play in the world ecosystem. It is common knowledge now that the forests are the producers of most of the oxygen that we consume. Photographs from Earth's orbit show the formation of rain clouds over the Amazonian rain forest. It is actually possible to see the forest system breathe.

The Earth system, Gaia, has regulated itself for millions of years. For instance, the oysters in Chesapeake Bay could at one time filter all the water in the bay in three days. Now, due to overharvesting and pollution, their numbers have fallen so low that WorldWatch researchers note it takes them one year to perform the same task.

A number of years ago it was discovered that plants grew better when exposed to music. Researchers, in attempting to discover just what kind of music produced the most growth, experimented with everything from classical music to Hindu ragas to rock and roll. A researcher, Dan Carlson, attempted to discover just why this was. Plants contain minute, mouthlike openings in their leaves called stomata. Stomata are used by plants to exchange gaseous aerosols and mists with the surrounding atmosphere. In other words, plants can feed off the atmosphere through the action of their stomata in the leaves. Carlson found that this action was increased by music. In working with a musician, Michael Holtz, they found that certain classical music was the most effective—most specifically *Spring* by Antonio

Lucio Vivaldi. The Vivaldi piece was extremely similar to the Bach violin sonatas broadcast by Ottawa researchers to a wheatfield. This had resulted in yields 66 percent greater than average.

Holtz took this even further when he realized that the violin passages in *Spring* were mimicry of birdsong. He began to examine the electronic frequencies and amplitudes of the music and birdsong. He discovered that birdsong was the perfect frequency and amplitude to cause stomata to increase their action to the highest level.[5] The singing of birds, developed over millennia in the ecosystem, helped plants to grow. The incredible die-off of bird species since the introduction of pesticides has had a tremendously negative impact on the forest systems. The bird species whose function, in part, is to help the trees and plants feed and form the atmosphere, are no longer there. Many are extinct, most others severely depleted in numbers.

As we humans turn to understanding how the ecosystem works, the focus is often on larger elements. Yet little is known about the function of plants in the ecosystem (even less about microbes) and the plants are dying off in unparalleled numbers.

The normal background rate of extinction of species on Earth is one to ten species a year. Generally, extinctions are of animal not plant species. Harvard University biologist Edward O. Wilson estimates that the current rate of extinction in the rain forests alone is 140 species per day, some 50,000 per year.[6] Some estimates put the die-off of species in rain forests at 15,000,000 species in the next hundred years,[7] many of them plant species. There have been large die-offs of animal species over our evolutionary history, but never has there been a die-off of plant species like the one now occurring. Plants have evolved for some reason. Given that form follows function, it would seem clear that plants fulfill some essential function in the world. To many, that function is unknown. But as Aldo Leopold noted in his seminal book, *A Sand County Almanac:*

If the biota, in the course of aeons, had built something we like but do not understand, then who but a fool would discard seemingly useless parts? To keep every cog and wheel is the first precaution of intelligent tinkering.[8]

On the land in Colorado where I lived for so long, and even here at my new home on Vision Mountain in northeastern Washington, the trees are in many instances ill and dying. One of the main causes now, the experts say, is dwarf mistletoe.

I have thought a lot about dwarf mistletoe. I rather energetically tried to root it out of our Colorado trees for five or six years, but all the time something seemed amiss with that approach. Then I began to pray with it and think on it and it slowly began to dawn on me what was happening. For in that time when usnea came to me and shared with me that it was an herb for the lung systems of the planet, I began to understand that the herbs perform regulatory functions in the ecosystem. I researched dwarf mistletoe and found that many indigenous cultures used it for hemorrhage from the lungs. I find it interesting in the extreme that usnea performs antibiotic function for lungs and that dwarf mistletoe is traditionally used when the lungs deteriorate further. I realized that these plants were helping to maintain the health of the trees. For the first time I had a direct experience of how Gaia regulates herself.

Finally, I began to understand that how herbs help us is only secondary to what they do for the Earth, for Gaia. Their primary function is in helping the health of the whole lovely life form that is Mother Earth.

This concept of Gaia has caught on strongly for many people. It is as if something we have always known has at last been said out loud. Vaclav Havel remarks on this when he notes that the Gaia Hypothesis reminds us

... in modern language, of what we have long suspected, of what we have long projected into our forgotten myths and what perhaps has always lain dormant within us as archetypes. That is, the awareness of our being anchored in the Earth and the Universe, the awareness that we are not here alone nor for ourselves alone, but that we are an integral part of higher, mysterious entities against whom it is not advisable to blaspheme. This forgotten awareness is encoded in all religions. All cultures anticipate it in various forms. It is one of the things that form the basis of man's understanding of himself, of his place in the world, and ultimately of the world as such.[9]

James Lovelock does not advocate that Gaia is a conscious being or aware, but simply that it is a living, holistic system. From my own experience and the research that I have done, I believe that Gaia, mother Earth, is purposeful and that the long established behavior patterns of indigenous peoples indicated a knowledge of this fact. They developed specific patterns of communication and relationship with other forces larger than themselves. These forces, which they called mother Earth and Creator, we now call Gaia and God. (Is it not an obvious thought that if the self regulating Earth is Gaia, by the same token the self regulating universe is God?) The patterns of communication that indigenous peoples created to communicate with these forces, they called ceremony. Of those patterns that were developed for relationship with Gaia, some are the ancient patterns of behavior that I call sacred plant medicine.

These ancient patterns represent basic articulations of communication with Gaia. We can use these patterns in the solutions to the many environmental crises facing our species.

When I teach I often work in offices or classrooms, and oftentimes I will ask my students at that time to touch the Earth nearest them. They most often respond with uncomprehending stares as they look around the man-made environment in which we are working. It is most uncommon that any of them will reach up and touch their own faces.

In the process of knowing the plants, of holding them in our hands, we come to learn in time that they hold us, too. And when we understand that, it is very easy to touch our own faces and in so doing, touch Gaia.

I believe that we are on the verge of a new time, that the knowledge of Gaia will gradually begin supplanting much of how we view ourselves as a species on the Earth. The innate understanding of the interconnectedness of all life forms that is an integral part of the Gaia hypothesis will become, in time, an integral aspect of how each of us approaches the Earth, each other, and ourselves. The prevalence of this new time can allow the reclaiming of older ways of gathering knowledge, in which sacred plant medicine can again become a common method of healing.

Within each of us is the possibility to restore our connections to this Earth, to hear the plants, and all life, speak. In that process we will truly know what to do, how to grow, how to change to embody the living Earth of which we are made, to know Gaia. And to some extent, the words of the old ones who have gone before us can point us in the right direction, can point to Gaia, if only we will listen. As the Teton Sioux, Oku'te, shared with Frances Densmore:

All living creatures and plants derive their life from the sun. If it were not for the sun, there would be darkness and nothing could grow—the earth would be without life. Yet the sun must have the help of the earth. If the sun alone were to act upon animals and plants, the heat would be so great that they would die, but there are clouds that bring rain, and the action of the sun and earth together supply the moisture that is needed for life. The roots of the plant go down, and the deeper they go the more moisture they find. This is according to the laws of nature and is one of the evidences of the wisdom of Wakan'tanka. Plants are sent by Wakan'tanka and come from the ground at his command, the part to be affected by the sun and rain appearing above the ground and the roots pressing downward to find the moisture which is supplied for them. Animals and plants are taught by Wakan'tanka what they are to do. Wakan'tanka teaches the birds to make nests, yet the nests of all birds are not alike. Wakan'tanka gives them merely the outline. Some make better nests than others. In the same way some animals are satisfied with very rough dwellings, while others make attractive places to live. Some animals also take better care of their young than others.

The forest is the home of many birds and other animals, and the water is the home of fish and reptiles. All birds, even those of the same species, are not alike, and it is the same with animals and with human beings. The reason Wakan'tanka does not make two birds, or animals, or human beings exactly alike is because each is placed here by Wakan'tanka to be an independent individuality and to rely on itself. Some animals are made to live in the ground. The stones and minerals are placed in the ground by Wakan'tanka, some stones being more exposed than others. When a medicine man says that he talks with the sacred stones, it is because of all the substances in the ground these are the ones that most often appear in dreams and are able to communicate with men.

All animals have not the same disposition. The horse, dog, bear, and buffalo all have their own characteristics. This is also true of the fowls of the air, the living creatures in the water, and even the insects, they all have their own ways. Thus a man may enjoy the singing of all the birds and yet have a preference for the melodies of certain kinds of birds. Or he may like all animals and yet have a favorite among them.

From my boyhood I have observed leaves, trees, and grass, and I have never found two alike. They may have a general likeness, but on examination I have found that they differ slightly. Plants are of different families, each being adapted to growth in a certain locality. It is the same with animals; they are widely scattered, and yet each will be found in a certain environment to which it is best adapted. It is the same with human beings, there is some place which is best adapted to each. The seeds of the plants are blown about by the wind until they reach the place where they will grow best—where the action of the sun and the presence of moisture are most favorable to them, and there they take root and grow. All living creatures and plants are a benefit to something. Certain animals fulfill their purpose by definite acts. The crows, buzzards, and flies are somewhat similar in their use, and even the snakes have a purpose in being. In the early days the animals probably roamed over a very wide country until they found their proper place. An animal depends a great deal on the natural conditions around it. If the buffalo were here today, I think they would be different from the buffalo of the old days because all the natural conditions have changed. They would not find the same food nor the same surroundings. We see the change in our ponies. In the old days they could stand great hardship and travel long distances without water. They lived on certain kinds of food and drank pure water. Now our horses require a mixture of food, they have less endurance and must have constant care. It is the same with the Indians; they have less freedom and they fall an easy prey to disease. In the old days they were rugged and healthy, drinking pure water and eating the meat of the buffalo, which had a wide range, not being shut up like the cattle of the present day. The water of the Missouri River is not pure, as it used to be, and many of the creeks are no longer good for us to drink.[10]

In spite of the difficulty of our times, in spite of the pain of the Earth, in spite of the difficult learnings before us as a species, I still believe in the capacity of *Homo sapiens* to adapt and to take their place in the council of life we call Gaia. Part *of*, not apart *from*.

Perhaps some of the knowledge of how to make that transition lies in listening to such humble things as the grasses beneath our feet, the rhythmic beating of our own hearts, and the insistent, loving urging of our mother, the Earth.

For the Earth

WILDCRAFTING ETHICS AND GUIDELINES
Prepared by the Rocky Mountain Herbalists' Coalition

Wildcrafting

1. Never gather an endangered or threatened species. Check your local herbarium or botanical garden for a list of these plants. You may also contact the American Herbalist Guild for a national listing: AHG, Box 1683, Soquel, CA 95073.

2. I.D. positively before harvesting. Use identification keys and voucher specimens.

3. Ask permission and give thanks, acknowledge connection with all life, share your appreciation.

4. Leave mature and seed-producing plants—grandparent plants—within the stand and at the top of a hill to seed downslope. Work your way up.

5. If unsure, harvest no more than ten percent native whole plant and root, and thirty percent naturalized plant species or native leaves and flowers. Gather only from abundant stands. Harvest conservatively to insure maintenance and well-being of plant communities.

Site Selection

1. Obtaining permission: On BLM land, a free use permit may be obtained for a minimal charge if you are collecting small amounts. Both the U.S. Forest Service and BLM will tell you there is no picking (a) in or near campgrounds or picnic areas; (b) any closer than 200 feet from trails; and (c) on the roadsides.

2. Stay away from downwind pollution, roadsides (at least 50 feet), high-tension electric wires (may cause mutations), fertilizers in lawns and public parks, downstream from mining or agribusiness, around parking lots, and possible sprayed areas. Some BLM and Forest Service districts use routine spraying. This applies to private land as well, and you may need to ask about herbicides and pesticides.

3. Use discretion with fragile environments—one irresponsible wildcrafter can severely alter a rocky hillside or streamside ecosystem.

Gardening and Propagation Techniques

1. Using proper wildcrafting techniques will insure minimal impact, increase harvest yields, and continue to provide plant food for wildlife. Do not harvest the same stand year after year, but tend the area as necessary. "Gardening" techniques that apply include thinning, root division, top pinching, and preserving a wide selection of grandparent plants to seed and guard young plants.

2. Be aware of erosion factors. *If digging roots, replant or scatter seeds, and cover holes.* Be mindful of hillside stands, replace foliage and dirt around harvested areas. Gathering foliage from nearby harvested plants and spreading it around may be necessary. Wearing hard-soled shoes may cause delicate hillside ecosystems irreparable damage.

3. If harvesting leaf, don't pull the roots. Flower pruning of certain plants will increase root yields as well as foliage.

4. Make seasonal observations on wildcrafted areas. Be mindful of

your harvested stands and check different growth cycles. This will determine your real impact on the ecosystem. (One experienced wildcrafter in the northwest has observed that a healthy population will increase about 30% a year until it reaches stasis. Anything less than this could be considered degenerative.)

Suggested Gathering Times

1. Aerial or above ground parts: Mornings between 6 and 10 a.m., just before they wilt in the sun. If harvesting leaf, many are best just before flowering. Harvest most flowers just as they are beginning to bloom—you should be able to see the color of the bud. The traditional moon phase for harvesting aerial parts is near or during the full moon.

2. Roots: Harvest after seeding; if possible, in the early morning before the sun hits. Biennials: Harvest in the autumn of the first year or the spring of the second year. The traditional phase is the new moon.

3. Barks: Harvest in the spring or fall. Never strip. Take the whole tree. Tree thinning is appropriate in dense populations, but always leave the healthiest looking trees. If you take from the small branches only, be aware of potentially leaving the tree vulnerable to fungal rot. For many barks, the inner bark, or cambium, is the most active. Leave short trunks for pollarding, and low stumps for coppicing. This will provide an ongoing harvest. The traditional phase for barks is the three-quarter waning moon.

4. Saps and Pitches: Harvest in late winter or early spring.

5. Seed and Fruit: Harvest when mature, with some exceptions such as citrus, unripe scarlet bean pods, etc.

Drying

1. Dry most plants in shaded, well ventilated areas, avoid wire screens and newspaper print. Research which plants dry better in the sun.

2. Don't wash leaf or flowers. Shake them to get bugs and dust off. If the quantities are manageable, tie bundles at the base of the stems in diameters of 1½ inches or less. They may also be scattered loosely on screens to dry.

3. Barks: Scrape off the outer bark if appropriate. This is called Tossing.

4. Roots: Lay them out or string. Rinsing usually will not remove soil particles. A pressure hose is often required as well as hand brushing, especially with clay. Cut lengthwise for large heavy roots without aromatic properties.

5. All plant parts are dry when brittle. Pinch the lower part of hanging plants. Cut large sample root in half to see if center is dry.

Storage

1. Avoid light and excessive heat that could destroy aromatic properties and other valuable constituents. When totally dry, food grade plastic bags or fiber barrels or other containers that omit oxygen and moisture are desirable to preserve quality and potency as long as possible.

2. Label with dates and location.

3. Broken or crushed herbs lose their value more rapidly than whole, uncut herbs.

The Rocky Mountain Herbalists' Coalition has developed approaches to the protection of wild medicinal plant populations. The motivation for creating these approaches comes from deep feelings for the sanctity of the Earth and plant populations. The approaches are designed to encompass enlightened self interest and social pressure as a nongovernmental organization active in a specific bioregion. These solutions are offered to those in other bioregions to be used, adapted and applied to their own particular needs.

The first step has been to create a set of guidelines to help harvesters understand the basic tenets of ethical wildcrafting. These guidelines have been developed with the input of many seasoned wildcrafters and subjected to extensive peer review. They are designed to supplement each wildcrafter's ethical philosophy, rather than to be hard and fast rules.

A criteria sheet providing information about where and how a plant is harvested, processed, and dried has been developed for the wildcrafter to fill out and send along with each shipment of herbs. This information provides buyers with an opportunity to become educated about ethical wildcrafting and actively support harvesters whose philosophies are in line with their own.

The next step was the creation of a Direct Marketing Registry of ethical wildcrafters and organic growers. The registry was designed to encourage responsible behavior by putting herb buyers in touch with the actual harvesters. This direct communication facilitates an environment of mutual respect, helps develop quality control, and allows wildcrafters to establish prices that support and encourage ethical harvesting practices.

Herbal product manufacturers are urged to use the registry and incorporate its principles into their daily business practices. Ecologically minded wildcrafters and organic growers are encouraged to submit additional listings to the current registry.

The latest step in this project is the creation of a bioregional cooperative of wildcrafters and organic growers. The idea is to come together as a community to exchange information; promote education among harvesters and growers; create structures for self regulation through peer review; and meet demands for ethically harvested and organically grown herbs at a price that supports sustainable, ecologically sound practices.

The co-op is a vehicle for members to share information on plant growth patterns, harvesting techniques, propagation and cultivation methods, as well as alternatives to over-harvesting. A list of native plants that are threatened, endangered, over harvested or improperly harvested, is being developed and limits set, via peer agreement, to insure species survival. This information can be shared with other wildcrafters, bioregional co-ops and buyers via a newsletter. A seed exchange program can also be developed so that areas can be reseeded for future use.

With the co-op, members will be able to share information on harvesting areas to ensure that certain areas are not being overused and that other areas are left alone to regenerate. The intention of the co-op is to create a structure that can be used to guide one another and cultivate personal integrity. The co-op can also be used as a place for people to come together as activists to protect places from logging, development, and grazing. We would like to see the creation of an annual gathering where wildcrafters can get together, share information, strengthen the network and become a vocal, visible part of the herbal community.

For those who hear the sound of the Earth crying, whether herbalist or not, there is an imperative to do something to respond. This is a beginning to protect this small portion of the Earth and help it continue. The Rocky Mountain Herbalists' Coalition believes that we must be mindful and not destroy wild plant populations or the wild places where they live, either directly through our own actions or indirectly through our tacit consent. To do this we must be careful to place long term sustainability before short term profits, on the grass roots level and the corporate level. If we collectively take a stand and adhere to ethical standards, those lacking respect for the plants will either change their harvesting techniques or fade from the market. In this way we strive to repair the damage done by our forebears, maintain our own sacred trust to the Earth, and give to our descendants the gift of reverence for life, demonstrated in a sustainable and healthy planet.

NOTES

Preface

1. *The Reader's Digest Great Encyclopedic Dictionary.* Pleasantville, N.Y., 1977.
2. Ibid.

Chapter 1. The Sacred and the Earth

1. Del Tredici, Peter. *Arnoldia,* 51(2):2–15, 1991, revised by the author as Classic Botanical Reprint Number 224, 1992.
2. I have heard various versions of this story over the years. Kelly Kindscher notes that Phillip Wells, a botany professor at the University of Kansas at Lawrence, tells a version of the story wherein the man who cut down the tree was a U.S. Forest Service employee and who, ironically, died not long afterward of a heart attack.
3. Leopold, Aldo. *A Sand County Almanac.* New York: Ballantine, 1991, p. 197.
4. Dudley, Guilford III. *Religion on Trial: Mircea Eliade and His Critics,* Philadelphia: Temple University Press, 1977, p. 50.
5. Underhill, Evelyn. *Mysticism.* New York: E. P. Dutton, 1961.
6. Otto, Rudolph. *The Idea of the Holy.* New York: Oxford University Press, 1978.
7. Yoon, Hong-key. *Geomantic Relationships Between Culture and Nature in Korea.* Chinese Association for Folklore, 1976.
8. Eliade, Mircea. *The Sacred and the Profane.* New York: Harcourt Brace, 1959.
9. Ryan, J. *Life Support: Conserving Biological Diversity.* WorldWatch Paper 108, April 1992, p. 17.

10. Ibid.
11. Radin, Paul. *Crashing Thunder: The Autobiography of an American Indian.* New York: Appleton, 1920.

Chapter 3. Visions of Sacred Plant Medicine

1. Dudley, Guilford III. *Religion on Trial.* p. 52.
2. Moerman, Daniel E. *Medicinal Plants of Native America.* University of Michigan Museum of Anthropology, Research Reports in Ethnobotany, Contribution 2, vol. 1, 1986
3. Lamb, F. Bruce. *Wizard of the Upper Amazon.* Boston: Houghton Mifflin, 1974.
4. Densmore, Frances. *Teton Sioux Music.* Smithsonian Institution: Bureau of American Ethnology Bulletin 61, p. 183, 1918.
5. Messer, Ellen. "Present and Future Prospects of Herbal Medicine in a Mexican Community," in *The Nature and Status of Ethnobotany,* by Richard Ford. Ann Arbor: University of Michigan Museum of Anthropology, Anthropology Paper 67, p. 146, 1798.
6. Stevenson, Matilda Coxe. *Ethnobotany of the Zuni Indians.* Smithsonian Institution: Bureau of American Ethnology Annual Report 30, p. 36, 1915.
7. Densmore, Frances. *Papago Music.* Smithsonian Institution: Bureau of American Ethnology Bulletin 90, p. 90, 1929.
8. Kindscher, Kelly. *Medicinal Plants of the Prairie.* Lawrence: University Press of Kansas, 1992.
9. Densmore, Frances. *Teton Sioux Music.* p. 251.
10. *The Sixth Grandfather: Black Elk's Teachings Given to John Neihardt.* Edited by Raymond J. DeMallie. Lincoln: University of Nebraska Press, p. 134, 1985.
11. Ibid., pp. 235–236.
12. Densmore, Frances. *Chippewa Music II.* Smithsonian Institution: Bureau of American Ethnology Bulletin 53, p. 37, 1913; *Menominee Music.* Bulletin 102, p. 76, 1934; *Uses of Plants by the Chippewa Indians.* Annual Report 44, p. 322, 1928; *Papago Music.* Bulletin 90, p. 82, 1929.
13. Sturtevant, W. C. *The Mikasuki Seminole: Medical Beliefs and Practices.* Ph.D. dissertation, Yale University, p. 143–146, 1955.
14. Densmore, Frances. *Nootka and Quileute Music.* Smithsonian Institution: Bureau of American Ethnology, Bulletin 124, p. 252, 1934.

Chapter 4. The Sacred Song of a Plant

1. Halifax, Joan. *Shamanic Voices.* New York: E. P. Dutton, p. 32, 1979.
2. Sturtevant, W. C. *The Mikasuki Seminole.* p. 155.

3. Swanton, John R. *Creek Religion and Medicine*. Smithsonian Institution: Bureau of American Ethnology Report 42, 1928.

4. Densmore, Frances. *Chippewa Music* II. p. 16.

5. Underhill, Evelyn. *Mysticism*.

6. Densmore, Frances. *The Belief of the Indian in a Connection Between Song and the Supernatural*. Smithsonian Institution: Bureau of American Ethnology Bulletin 151, Anthropology Paper 37, p. 217–223, 1953.

7. Ibid.

8. Fenton, William. *Iroquois Suicide*. Smithsonian Institution: Bureau of American Ethnology Bulletin 128, p. 86, 1941.

9. Ibid., p. 88.

10. Harrison, Kathleen. *Seeds of Change Catalog*. p. 67, 1994.

11. Densmore, Frances. "Music in the treatment of the sick," in *Frances Densmore and American Indian Music: A Memorial Volume*. Charles Hofmann, editor. New York: Heye Foundation, Museum of the American Indian, p. 72, 1968.

12. Halifax, Joan. *Shamanic Voices*.

13. Densmore, Frances. *Teton Sioux Music*. p. 254.

14. Ibid., p. 255.

15. Ibid., p. 260.

16. Ibid., p. 262–263.

17. Mooney and Olbrechts. *The Swimmer Manuscript: Cherokee Sacred Formulas and Medicinal Prescriptions*. Smithsonian Institution: Bureau of American Ethnology Bulletin 59, p. 156, 1932.

Chapter 5. Making Sacred Relationship with Plants

1. Fukuoka, Masanobu. *The Natural Way of Farming*. Tokyo: Japan Publications, 1985.

2. Nabhan, Gary Paul. *Enduring Seeds*. North Point Press, 1989.

3. Densmore, Frances. *Uses of Plants By the Chippewa Indians*. p. 325.

4. Fukuoka, Masanobu. *The Natural Way of Farming*.

5. Wall, Steve, and Harvey Arden. *Wisdomkeepers*. Beyond Words Publishing, p. 80, 1990.

6. Conklin, Harold. *The Relation of Hanunoo Culture to the Plant World*. Ph.D. dissertation, Yale University, 1955.

7. Carneiro, Robert. "The Knowledge of Forest and Rain Forest Trees by the Kuikuru Indians of Central Brazil," in *The Nature and Status of Ethnobotany*, by Richard Ford. Ann Arbor: University of Michigan Museum of Anthropology, Anthropology Paper 67, p. 205.

8. Robbins, W. W., J. P. Harrington, and B. Freire-Marreco. *Ethnobotany of the Tewa Indians.* Smithsonian Institution: Bureau of American Ethnology Bulletin, 55, p. 9, 1916.

9. Plotkin, Mark. "The Healing Forest," in *The Futurist.* Jan/Feb 1990.

10. Ford, Richard. *The Nature and Status of Ethnobotany.* p. 43.

11. Messer, Ellen. *Zapotec Plant Knowledge: Classification, Uses, and Communication about Plants in Mitla, Oaxaca, Mexico.* Ann Arbor: University of Michigan Museum of Anthropology Memoirs 10, part 2, p. 69, 1978.

12. Hoffman, W. J. *The Mide'wiwin or "Grand Medicine Society" of the Ojibwa.* Smithsonian Institution: Bureau of American Ethnology Bulletin, Annual Report 7, 1891.

13. Herrick, James William. *Iroquois Medical Botany.* Ph.D. dissertation, State University of New York at Albany, p. 136, 1977, quoting David Jack as recorded by Waugh.

14. Conklin, Harold. *The Relation of the Hanunoo Culture to the Plant World.* p. 239.

15. Fukuoka, Masanobu. *The Natural Way of Farming.* p. 17.

16. Herrick, James William. *Iroquois Medical Botany.* p. 136–137, quoting Jesse Cornplanter, as recorded by Fenton.

17. Harrington, John. *Tobacco Use Among the Karuk Indians of California.* Smithsonian Institution: Bureau of American Ethnology Bulletin 94, 1932.

18. Densmore, Frances. *Teton Sioux Music.* p. 254.

19. Mooney and Olbrechts, *The Swimmer Manuscript.*

Chapter 6. Healing in a Sacred Manner

1. Densmore, Frances. *Teton Sioux Music.* p. 245.

2. Sturgeon, Theodore. "Scars," in *E Pluribus Unicorn.* New York: Pocket Books, 1977.

3. Eliade, Mircea. *Shamanism.* Princeton: Princeton University Press, p. 229, 1972.

4. Densmore, Frances. *Chippewa Music I.* Smithsonian Institution: Bureau of American Ethnology Bulletin 45, p. 119, 1910.

5. Swanton, John R. *Beliefs and Usages of the Chickasaw.* Smithsonian Institution: Bureau of American Ethnology Annual Report 44, p. 268, 1928.

6. Dudley, Guilford. *Religion on Trial.* p. 77.

7. Meyerhoff, Barbara. "Shamanic Equilibrium: Balance and Mediation in Known and Unknown Worlds," in *American Folk Medicine,* Wayland Hand, editor. Berkeley: University of California Press, p. 103, 1976.

8. Ibid., p. 101.
9. Canda, Edward. Personal communication, 1993. See also Edward Canda. "Gripped by the Drum: The Korean Tradition of Nongak," in *Shaman's Drum* 33, Fall/Winter 1993.
10. Densmore, Frances. *Papago Music.* p. 116.
11. Popov, Andrei. "How Sereptie Djarroskin of the Nganasans (Tavgi Samoyeds) Became a Shaman," in *Popular Beliefs and Folklore Traditions in Siberia,* Vilmos Dioszegi, editor. Translated by Stephen Dunn. Bloomington: Indiana University Press, p. 137–146, 1968.
12. Furst, Peter. "Huichol Conception of the Soul," in *Folklore Americas.* vol. 27, No. 2, p. 52–56, June 1967.
13. Eliade, Mircea. *Shamanism.* Princeton: Princeton University Press, 1972.
14. Grof, Stanislav, and Joan Halifax. *The Human Encounter with Death.* New York: E. P. Dutton, 1978.
15. Eliade, Mircea. *Shamanism.* p. 248.
16. Swanton, John R. *Creek Religion and Medicine.* p. 638. See also Joseph Bruchac. *The Native American Sweatlodge.* Freedom, Calif.: Crossing Press, 1993.
17. Mooney, James. *Sacred Formulas of the Cherokee.* Smithsonian Institution: Bureau of American Ethnology Annual Report 7, p. 322, 1891.
18. Ibid.
19. Fenton, William N. *Iroquois Suicide.* p. 86.

Chapter 7. Digging for Medicine:
The Wildcrafting of Medicinal Plants

1. Densmore, Frances. *Menominee Music.* p. 119.
2. Leopold, Aldo. *A Sand County Almanac.* New York: Sierra Club Books, pp. 138–39, 1966.
3. Densmore, Frances. *Teton Sioux Music.* 1918.
4. Sturtevant, W. C. *The Mikasuki Seminole.* 1955.
5. Moore, Michael. *Medicinal Plants of the Mountain West.* Sante Fe: Museum of New Mexico Press, 1979.
6. Densmore, Frances. *Menominee Music.* p. 118–19.

Chapter 8. Making Plants into Medicine:
The Technology of Herbalism

1. Densmore, Frances. *Chippewa Music II.* p. 65.
2. Hoffmann, David. "Deep Ecology and Deep Herbalism," in *Plants of Power: Earthkeepers of Six Nations Speak on the Sacred Power of Plants,* unpublished manuscript by Stephen Buhner.

3. Bishop, Jerry. "Research on Duodenal Ulcer Concludes Eradication of Bacteria is Cost Effective," in *The Wall Street Journal*. November 1, 1995.
4. Farnsworth, Norman. *The Present and Future of Pharmacognosy*. American Botanical Council reprint 209, reprinted from *The American Journal of Pharmaceutical Education*, 43:239–243 (1979).
5. Densmore, Frances. *Uses of Plants by the Chippewa Indians*. p. 323.

Chapter 9. Ceremony:
Making Deeper Relationship with Plants

1. Freesoul, Jonathon. *Breath of the Invisible*. New York: Theosophical Publishing, 1986.
2. Catlin, George. *The Catlin Manuscript*. Washington: British Museum Publications and Smithsonian Institution Press, 1979.
3. Brown, Joseph Epes. *The Sacred Pipe*. Norman: University of Oklahoma Press, p. 5–7, 1989.
4. Hofman, Charles. *Frances Densmore and American Indian Music*. New York: Museum of the American Indian, Heye Foundation, Vol. 23, 1968.
5. Fortune, R. F. *Omaha Secret Societies*. Columbia University Contributions to Anthropology, vol. 14: Columbia University Press, 1932. Densmore, Frances. *Teton Sioux Music; Chippewa Music; Mandan and Hidatsa Music*, 1923.
6. Brown, Joseph Epes. *The Sacred Pipe: Black Elk's Account of the Seven Rites of the Oglala Sioux*. New York: Penguin Books, 1981.
7. Ibid.
8. Powers, William K. *Yuwipi: Vision and Experience in Oglala Ritual*. Lincoln: University of Nebraska Press, p. 24–25, 1982.

Chapter 10. Four Sacred Plants of the Rocky Mountain Region:
Their Sacred Qualities and Uses as Medicine

1. Kindscher, Kelly. *Medicinal Plants of the Prairie*.
2. Felter and Lloyd. *King's American Dispensatory*. Sandy: Eclectic Medical Publications, 1993.
3. Gilmore, Melvin. *Uses of Plants by the Indians of the Missouri River Region*. University of Nebraska Press, p. 29, 1991.
4. Curtin, L. S. M. *Healing Herbs of the Upper Rio Grande*. Los Angeles: Southwest Museum, 1965; Moore, Michael. *Medicinal Plants of the Mountain West*. Sante Fe: Museum of New Mexico Press, 1979.
5. Densmore, Frances. *Teton Sioux Music*. p. 195.
6. Ibid., p. 196.

7. Hale, Mason. *The Biology of Lichens.* New York: American Elsevier Publishing Co., Inc. 1974; Hobbs, Christopher. *Usnea: The Herbal Antibiotic.* Capitola, Calif.: Botanica Press, 1990.

8. Hobbs. *Usnea.*

9. Ahmadjian and Hale. *The Lichens.* New York: Academic Press, 1973.

10. *Napralert.* "Ethnomedical Information on Usnea, Including Biomedical Activities and Compound Presence," 1991.

11. Felter and Lloyd. *King's American Dispensatory.* Hoffmann, David. *The Herbal Handbook.* Rochester: Healing Arts Press, 1988.

12. Moerman, Daniel E. *Medicinal Plants of Native America.* Ann Arbor: University of Michigan Museum of Anthropology, Research Reports in Ethnobotany, Contribution 2, 1986.

Chapter 11. A Short Compendium of Plants and Their Sacred Uses

1. Felter and Lloyd. *King's American Dispensatoiy.* Hoffmann, David. *The Herbal Handbook.*

2. Boas, Franz. *The Religion of the Kwakiutl Indians.* New York: Columbia University Press, p. 212–13, 1930.

3. Felter and Lloyd. *King's American Dispensatory.*

4. Hoffmann, David. *The New Holistic Herbal.* Rockport, Mass.: Element, 1992.

5. Werbach, Melvyn, and Michael Murray. *Botanical Influences on Illness.* Tarzana, Calif.: Third Line Press, 1994.

6. Heron, Selena. "Botanical Treatment of Chronic Gynecological Conditions: Infertility, Endometriosis, and Symptoms of Menopause," in *American Herbalism.* Michael Tierra, editor. Freedom: Crossing Press, 1992.

7. Mowrey, Daniel B. *The Scientific Validation of Herbal Medicine.* New Canaan, Conn.: Keats Publishing, 1986.

8. Densmore, Frances. *Teton Sioux Music.* p. 164.

9. Moore, Michael. *Medicinal Plants of the Mountain West.*

10. Speck, Frank. *Catawba Texts.* New York: Columbia University Press, p. 54 and 60, 1934.

11. Drum, Ryan. "Joining the Devils Club," in *Plants of Power: Earthkeepers of Six Nations Speak on the Sacred Power of Plants* by Stephen Buhner, unpublished manuscript; Moore, Michael. *Medicinal Plants of the Pacific West.*

12. Boas, Franz. *The Religion of the Kivakiuti Indians.* p. 241.

13. Felter and Lloyd. *King's American Dispensatory;* Foster and Duke. *Eastern/Central Medicinal Plants.* Boston: Houghton Mifflin, 1990.

14. Densmore, Frances. *Chippewa Music II.* p. 63.
15. Ibid., p. 99.
16. Moore, Michael. *Medicinal Plants of the Desert and Canyon West;* Hobbs, Christopher. *The Echinacea Handbook.* Portland: Eclectic Medical Publications, 1989.
17. Densmore, Frances. *Teton Sioux Music.* p. 268.
18. Felter and Lloyd. *King's American Dispensatory and Medicology.* New York: University Medical Publications, 1906.
19. Goddard, Pliny Earle. *Hupa Texts.* Berkeley: University of California Publications, American Archaeology and Ethnology, vols. 1, 2. Berkeley: University Press, p. 276–77, 1903.
20. Moore, Michael. *Medicinal Plants of the Mountain West;* Harrington, H. D. *Western Edible Plants.* Albuquerque: University of New Mexico Press, 1972.
21. Speck, Frank. *Catawba Texts.* p. 55.
22. Felter and Lloyd. *King's American Dispensatory;* Foster and Duke. *Eastern/Central Medicinal Plants;* Moore, Michael. *Medicinal Plants of the Mountain West.*
23. Felter and Lloyd. *King's American Dispensatory;* Foster and Duke. *Eastern/Central Medicinal Plants;* Moore, Michael. *Medicinal Plants of the Mountain West.*
24. Densmore, Frances. *Teton Sioux Music.* p. 254–56.
25. Felter and Lloyd. *King's American Dispensatory;* Hutchins, Alma. *Indian Herbology of North America.* Windsor, Ontario: Merco, 1989.
26. Boas, Franz. *The Religion of the Kwakiutl Indians.* p. 237–38.
27. Moore, Michael. *Medicinal Plants of the Mountain West.*
28. Densmore, Frances. *Teton Sioux Music.* p. 259.
29. Felter and Lloyd. *King's American Dispensatory;* Foster and Duke. *Eastern/Central Medicinal Plants;* Mowrey, Daniel B. *The Scientific Validation of Herbal Medicine.* New Canaan: Keats Publishing, 1986; Hutchins, Alma. *Indian Herbology of North America;* Werbach, Melvyn, and Michael Murray. *Botanical Influences on Illness.* Tarzana, Calif.: Third Line Press, 1994.
30. Hoffmann, David. *The New Holistic Herbal;* Hutchins, Alma. *Indian Herbology of North America.*
31. Boas, Franz. *The Religion of the Kwakiutl Indians.* p. 221.

Chapter 12. The Birth of Gaia

1. Havel, Vaclav. Address of the President of the Czech Republic. His Excellency Vaclav Havel, on the occasion of the Liberty Medal Ceremony, July 4, 1994.

2. Lovelock, James. *Healing Gaia.* New York: Harmony Books, p. 11, 1991.

3. Mollison, Bill. *Permaculture: A Practical Guide for a Sustainable Future.* Washington, D.C.: Island Press, 1990.

4. Ibid.

5. Tompkins, Peter, and Christopher Bird. *Secrets of the Soil.* New York: Harper and Row, 1990.

6. Ryan, John. *Life Support: Conserving Biological Diversity.* WorldWatch Paper 108, 1992; Weber, Peter. "A Place for Pesticides?" in *WorldWatch Magazine,* vol. 5, no. 3, May/June 1992.

7. Diamond, Jared. "Playing Dice with Megadeath," in *Discover,* April, 1990.

8. Leopold, Aldo. *A Sand County Almanac.*

9. Havel, Vaclav. Address of the President of the Czech Republic.

10. Densmore, Frances; *Teton Sioux Music.*

SUGGESTED READINGS AND HERBAL APPRENTICESHIP PROGRAMS

Other approaches to sacred plant medicine can be found in the works of Machelle Small Wright and Susun Weed. Wright's work, which she calls Perelandra is similar to (in my opinion) Celtic work. Weed's work, which she calls "the Wise Woman tradition," is similar to (in my opinion) Wiccan work. The references cited also contain many books with a great deal of information on sacred plant medicine and Earth-centered spirituality. The ones listed here are the ones that deal with them most directly and clearly.

Sacred Plant Medicine

Frances Densmore. *Teton Sioux Music*. Bureau of American Ethnology Bulletin 61, 1918.

Machelle Small Wright. *Perelandra Flower Essenses*. Perelandra, 1980.

Susun S. Weed. *Wise Woman Herbal: Healing Wise*. Ash Tree Publishing, 1989.

Making Herbal Medicine and Herbalism

James Green. *The Herbal Medicine Maker's Guide*. Crossing Press, 2000. This is the best guide for making herbal medicines on the market. Its format and content are consistent with sacred plant medicine.

David Hoffmann. *The New Holistic Herbal*. Element, 1990. Overall the best herbal medicine book on the market. Written by someone who lived at the Findhorn community in Scotland and has incorporated that manner of relationship with the Earth.

Michael Moore. Any of his works but especially useful are his *Herbal Materia Medica* and *Herbal Repertory in Clinical Practice*. Both are available from the Southwest School of Botanical Medicine, www.swsbm.com.

Earth-Centered Relationship and the State of Mind Necessary for Sacred Plant Medicine

Joan Halifax. *Shamanic Voices*. Dutton, 1979.

F. Bruce Lamb. *Wizard of the Upper Amazon*. Houghton Mifflin, 1974.

Barbara Meyerhoff. *Peyote Hunt*. Cornell University Press, 1974.

The Sixth Grandfather: Black Elk's Teachings Given to John Neihardt. Edited by Raymond DeMallie. University of Nebraska Press, 1984.

James Lovelock. *Healing Gaia*, NY: Harmony Books, 1991.

Aldo Leopold. *A Sand County Almanac*. Oxford University Press, 1966.

Masanobu Fukuoka. *The Natural Way of Farming*. Japan Publications, 1985.

Bill Mollison. *Permaculture: A Designer's Manual*. Island Press, 1990.

Sun Bear. *Path of Power*. Bear Tribe Publishing, 1983.

Evelyn Eaton. *I Send a Voice*. Theosophical Publishing, 1978.

Brooke Medicine Eagle. *Buffalo Woman Comes Singing*. Ballantine Books, 1992.

Forest Carter. *The Education of Little Tree*. University of New Mexico Press, 1976.

Sacred Plant Medicine Apprentice Programs

Sacred Plant Medicine Apprentice Program
Stephen Buhner
www.gaianstudies.org

Wise Woman Apprentice Programs
Susun Weed
P.O. Box 64
Woodstock, New York 12498

Perelandra Apprentice Program
Machelle Small Wright
Perelandra
P.O. Box 3603
Warrenton, Virginia 22186
(540) 937-2153

INDEX

About the Author

Stephen Harrod Buhner is recognized by his contemporaries as being one of the most powerful and moving presenters in the country on biognosis, the deep ecology of herbalism, and the states of mind necessary for successful habitation of Earth. He is a specialist in cross-cultural religion, indigenous cosmologies, leading edge psychotherapies, and herbal medicine. His work has been profiled in the *New York Times,* on CNN and on Good Morning America.

The following publishers have generously given permission to use extended quotations from copyrighted works:

From *The Sixth Grandfather*, edited by Raymond J. DeMallie. Copyright 1984 by University of Nebraska Press. Reprinted by permission of the publisher. From *American Folk Medicine*, by Wayland Hand. Copyright 1976 by the Regents of the University of California. Reprinted by permission of University of California Press. From *The Sacred Pipe*, recorded and edited by Joseph Epes Brown. Copyright 1953, 1989 by University of Oklahoma Press. From YUWIPI, by William K. Powers. Copyright 1982 by University of Nebraska Press. Reprinted by permission of the publisher. From *Breath of the Invisible*, by John Redtail Freesoul. Copyright 1986 by John Redtail Freesoul. Reprinted by permission of the author. From *The Religion of the Kwakiutl Indians*, by Franz Boas. Copyright 1930 by Columbia University Press. Reprinted by permission of the publisher. From *Wildcrafting Ethics and Guidelines*. Copyright 1986 by the Rocky Mountain Herbalists' Coalition. Reprinted by permission of the Rocky Mountain Herbalists' Coalition.

Permission for use of the following photographs is gratefully acknowledged:

Photo of Ramon Medina Silva from *Peyote Hunt*, by Barbara C. Meyerhoff. Copyright 1974 by Cornell University Press. Used by permission of the publisher. Photo of shaman dancing on straw-chopping blades, by Edward R. Canda. Copyright 1977 by Edward R. Canda. Used by permission of the photographer. From *Arnoldia*. Photo of ginkgo tree in temple, by Stephen A. Spongberg, from *Arnoldia* 51:2. Used by permission of the Arnold Arboretum of Harvard University.

Books of Related Interest

The Secret Teachings of Plants
The Intelligence of the Heart in the Direct Perception of Nature
by Stephen Harrod Buhner

Medicine of the Cherokee
The Way of Right Relationship
by J. T. Garrett and Michael Tlanusta Garrett

The Cherokee Herbal
Native Plant Medicine from the Four Directions
by J. T. Garrett

Walking on the Wind
Cherokee Teachings for Harmony and Balance
by Michael Garrett

The Rebirth of Nature
The Greening of Science and God
by Rupert Sheldrake

The Universe Is a Green Dragon
A Cosmic Creation Story
by Brian Swimme, Ph.D.

Science and the Akashic Field
An Integral Theory of Everything
by Ervin Laszlo

The Encyclopedia of Psychoactive Plants
Ethnopharmacology and Its Applications
by Christian Rätsch

Inner Traditions • Bear & Company
P.O. Box 388
Rochester, VT 05767
1-800-246-8648
www.InnerTraditions.com

Or contact your local bookseller